Collins

Pupil Book 2.2

Maths Frameworking

3rd edition

Kevin Evans, Keith Gordon,
Trevor Senior, Brian Speed,
Chris Pearce

Contents

How to use this book

Learning objectives

See what you are going to cover and what you should already know at the start of each chapter.

About this chapter

Find out the history of the maths you are going to learn and how it is used in real-life contexts.

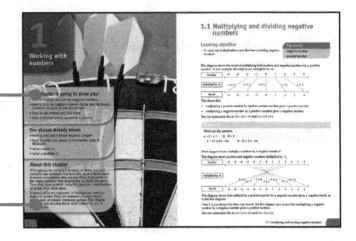

Key words

The main terms used are listed at the start of each topic and highlighted in the text the first time they come up, helping you to master the terminology you need to express yourself fluently about maths. Definitions are provided in the glossary at the back of the book.

Worked examples

Understand the topic before you start the exercises, by reading the examples in blue boxes. These take you through how to answer a question step by step.

Skills focus

Practise your problem-solving, mathematical reasoning and financial skills.

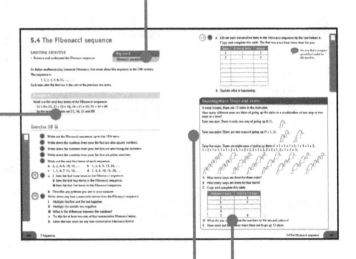

Take it further

Stretch your thinking by working through the **Investigation**, **Problem solving**, **Challenge** and **Activity** sections. By tackling these you are working at a higher level.

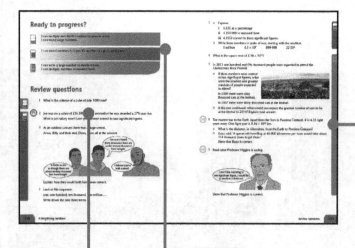

Progress indicators

Track your progress with indicators that show the difficulty level of each question.

Ready to progress?

Check whether you have achieved the expected level of progress in each chapter. The statements show you what you need to know and how you can improve.

Review questions

The review questions bring together what you've learnt in this and earlier chapters, helping you to develop your mathematical fluency.

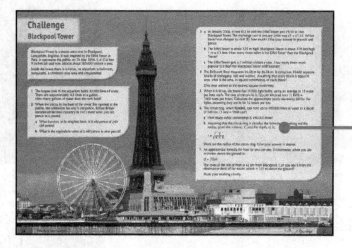

Activity pages

Put maths into context with these colourful pages showing real-world situations involving maths. You are practising your problem-solving, reasoning and financial skills.

Interactive book, digital resources and videos

A digital version of this Pupil Book is available, with interactive classroom and homework activities, assessments, worked examples and tools that have been specially developed to help you improve your maths skills. Also included are engaging video clips that explain essential concepts, and exciting real-life videos and images that bring to life the awe and wonder of maths.

Find out more at www.collins.co.uk/connect

1

Working with numbers

This chapter is going to show you:
- how to multiply and divide negative numbers
- how to find the highest common factor and the lowest common multiple of sets of numbers
- how to use powers and find roots
- how to find the prime factors of a number.

You should already know:
- how to add and subtract negative integers
- how to order operations, following the rules of BIDMAS
- what a factor is
- what a multiple is.

About this chapter

What games do you play? In many of them, you will certainly use numbers. For example, darts players need to make calculations very quickly. They must work out the target numbers they must score, to finish the game. Then they have to think about the possible combinations of scores from three darts.

Number skills are important in field games, such as rugby or cricket. They are essential in many board games and, of course, computer games. This chapter will help you develop those skills further for use in everyday life.

1.1 Multiplying and dividing negative numbers

Learning objective

- To carry out multiplications and divisions involving negative numbers

Key words

negative number

positive number

This diagram shows the result of multiplying both positive and **negative numbers** by a **positive number**. In this example all numbers are multiplied by +2.

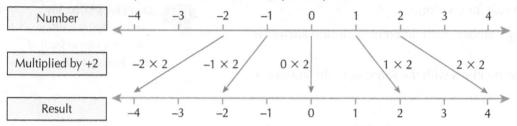

This shows that:

- multiplying a positive number by another positive number gives a positive number
- multiplying a negative number by a positive mumber gives a negative number.

You can summarise this as $(+) \times (+) = (+)$ and $(-) \times (+) = (-)$.

Example 1

Work out the answers.

a -11×4 **b** -8×3

    a $-11 \times 4 = -44$ **b** $-8 \times 3 = -24$

What happens if you multiply a number by a negative number?

This diagram shows positive and negative numbers multiplied by −2.

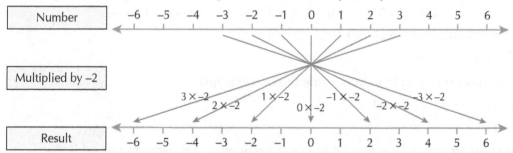

This diagram shows that multiplying a positive number by a negative number gives a negative result, as in the first diagram.

Here it is just shown the other way round. But this diagram also shows that multiplying a negative number by a negative number gives a positive number.

You can summarise this as $(-) \times (-) = (+)$ and $(+) \times (-) = (-)$.

To help you remember....

- When multiplying numbers with different signs, the answer is negative.
- When multiplying numbers with the same sign, the answer is positive.

Example 2

Work out the answers.

a $7 \times (-5)$ **b** -5×-8

   a $7 \times (-5) = -35$ **b** $-5 \times -8 = 40$

You use the same rule for division…

Hint   Don't forget that any number that is written without a sign in front of it is always positive.

- When dividing numbers with different signs, the answer is negative.
- When dividing numbers with the same sign, the answer is positive.

Example 3

Work out the answers.   **a** $-15 \div -5$ **b** $4 \div -2$ **c** $3 \times -8 \div -2$

   a $15 \div 5 = 3$ The signs are the same so the answer is positive.

     $-15 \div -5 = 3$

   b $4 \div 2 = 2$ The signs are different so the answer is negative.

     $4 \div -2 = -2$

   c $3 \times -8 = -24$

     $-24 \div -2 = 12$

     $3 \times -8 \div -2 = 12$

Example 4

Work these out.   **a** $-3 \times -2 - 5$ **b** $-3 \times (-2 - 5)$ **c** $-3 - 12 \div -2$

   a Using the rules of BIDMAS, work out -3×-2 first.

      $-3 \times -2 - 5 = 6 - 5 = 1$

   b This time you must do the calculation inside the brackets first.

      $-3 \times (-2 - 5) = -3 \times -7 = 21$

   c $-3 - 12 \div -2$

     It is important to recognise here that 12 is a positive number, since the minus (–) is the operation and not the sign of the number 12.

     So $-3 - 12 \div -2 = -3 - (12 \div -2) = -3 - -6 = -3 + 6 = 3$.

Exercise 1A

1 Work these out.

a $-7 + 8$ **b** $-2 - 7$ **c** $6 - 2 + 3$ **d** $-6 - 1 + 7$ **e** $-3 + 4 - 9$

f $-3 - 7$ **g** $-4 + -6$ **h** $7 - 6$ **i** $-3 - 7 + -8$ **j** $-5 + -4 - -7$

2 Copy these number walls, then subtract the right-hand number from the left-hand number to find the number in the brick below, each time. Some bricks have been completed for you.

a

7	-2	4	-1	5

9			-6

48

b

6	2	-3	4	-1

4			5

-37

3 Work out the answers.

a 2×-3 **b** -3×4 **c** -5×2 **d** -6×-3

e -3×8 **f** -4×5 **g** -3×-4 **h** -6×-1

i 7×-2 **j** -2×8 **k** -6×-10 **l** 8×-4

m -15×-2 **n** $-6 \times -3 \times -1$ **o** $-2 \times 4 \times -2$ **p** -7×-8

4 The answer to the question on this blackboard is -12.

Using multiplication and/or division signs, write down at least five different calculations that give this answer.

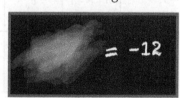

5 Work out the answers.

a $12 \div -3$ **b** $-24 \div 4$ **c** $-6 \div 2$ **d** $-6 \div -3$

e $-32 \div 8$ **f** $-40 \div 5$ **g** $-32 \div -4$ **h** $-6 \div -1$

i $7 \div -2$ **j** $12 \div 6$ **k** $60 \div -10$ **l** $+8 \div 4$

m $-15 \div -2$ **n** $-6 \times -3 \div -2$ **o** $-2 \times 6 \div -3$ **p** $9 \div -2$

6 Copy and complete each multiplication grid.

a

×	-2	3	-4	5
-3	6			
6				
-2				
5				

b

×	-1	-3	4	
-2		6		
		12		
			-5	
7			-42	

c

×				-8
-2		-12		
	-15		21	
4			28	
		-30		

7 Find the missing number in each calculation.

a $2 \times -3 = \square$ b $-2 \times \square = -8$ c $3 \times \square = -9$

d $\square \div -5 = -15$ e $-4 \times -6 = \square$ f $-3 \times \square = -24$

g $-64 \div \square = 32$ h $\square \times 6 = 36$ i $-2 \times 3 = \square$

j $\square \times -6 = -48$ k $-2 \times \square \times 3 = 12$ l $\square \div -4 = 2$

m $5 \times 4 \div \square = -10$ n $-5 \times \square \div -2 = -10$ o $\square \times -4 \div -2 = 14$

8 a Work these out.

     i -2×-2 ii -4×-4 iii $(-3)^2$ iv $(-6)^2$

     b Explain why it is impossible to get a negative answer when you square any number.

9 Work these out.

a $2 \times -3 + 4$ b $2 \times (-3 + 4)$ c $-2 + 3 \times -4$ d $(-2 + 3) \times -4$

e $-5 \times -4 + 6$ f $-5 \times (-4 + 6)$ g $-12 \div -6 + 2$ h $-12 \div (-6 + 2)$

10 Put brackets in each of these statements to make them true.

a $2 \times -5 + 4 = -2$ b $-2 + -6 \times 3 = -24$ c $9 - 5 - 2 = 6$

Challenge: Algebraic magic squares

This is an algebraic magic square.

A What is the 'magic expression' that is the total for every row, column and diagonal?

B Find the value in each cell, when $a = 7$, $b = 9$, $c = 2$.

C Find the value in each cell, when $a = -1$, $b = -3$, $c = -5$.

$a + c$	$c - a - b$	$b + c$
$b + c - a$	c	$a + c - b$
$c - b$	$a + b + c$	$c - a$

1.2 Factors and highest common factors (HCF)

Learning objective

- To understand and use highest common factors

Remember that the **factors** of an **integer** are the integers that divide exactly into it without leaving a remainder (or giving a decimal number as the answer). An integer is a whole number, whether it is positive or negative.

Key words

common factor

divisible

factor

highest common factor (HCF)

integer

Look at these examples.

- The factors of 6 are 1, 2, 3 and 6. The numbers 4 and 5 are not factors of 6 because if you divided them into 6 you would get a remainder (or a decimal answer).

- The factors of 25 are 1, 5 and 25. No other integers divide into 25 exactly.

It is important to remember that every integer (apart from 1) has at least two factors, 1 and itself. For example:

- $1 \times 17 = 17$

- The factors of 17 are 1 and 17.

Another way of saying 'can be divided by' is 'is **divisible** by'. The number 6 is divisible by the integers 1, 2, 3 and 6, and the number 25 is divisible by 1, 5 and 25. Every number is divisible by its factors.

Sets of numbers always have **common factors**. These are numbers that will divide into all of them. For example, the factors of 10 are 1, 2 and 5 and the factors of 15 are 1, 3 and 5. So 5 is a common factor of 10 and 15.

Some pairs of numbers, such as 2 and 3, or 4 and 5, only have 1 as a common factor.

Common factors can help you to solve some mathematical problems, such as simplifying fractions. If numbers have more than one common factor you should use the highest one, which is called the **highest common factor (HCF)**. Using this helps you to simplify a fraction as far as you can.

Example 5

Mr Bishop is taking two classes on a school visit. There are 27 pupils in one class and there are 18 pupils in the other. He wants to split the classes into smaller groups to be looked after by helpers. He wants all the groups to be the same size but does not want to mix pupils from different classes in any group. What is the largest size group he can choose, to split the classes, so that each group is the same size?

He can split the class with 27 pupils into groups of 1, 3, 9 and 27 (these are the factors of 27).

He can split the class with 18 pupils into groups of 1, 2, 3, 6, 9 and 18 (these are the factors of 18).

There are three common factors, 1, 3 and 9. The highest (HCF) is 9 so the largest groups of the same size Mr Bishop can create are groups of 9.

Example 6

Find the highest common factor (HCF) of the numbers in each pair.

a 15 and 21 **b** 16 and 24

  a Write out the factors of each number.

      15: 1, 3, 5, 15

      21: 1, 3, 7, 21

  You can see that the HCF of 15 and 21 is 3.

  b Write out the factors of each number.

      16: 1, 2, 4, 8, 16

      24: 1, 2, 3, 4, 6, 8, 12, 24

  You can see that the HCF of 16 and 24 is 8.

Example 7

Simplify these fractions.   **a** $\frac{18}{30}$ **b** $\frac{15}{27}$

To simplify fractions, you need to divide the numerator (top) and denominator (bottom) by their HCF.

a To simplify $\frac{18}{30}$:   the factors of 18 are 1, 2, 3, 6, 9, 18

the factors of 30 are 1, 2, 3, 5, 6, 10, 15, 30.

The HCF is 6, so divide the numerator and denominator of $\frac{18}{30}$ by 6 to give $\frac{3}{5}$.

b To simplify $\frac{15}{27}$:   the factors of 15 are 1, 3, 5, 15

the factors of 27 are 1, 3, 9, 27.

The HCF is 3, so divide the numerator and denominator of $\frac{15}{27}$ by 3 to give $\frac{5}{9}$.

Exercise 1B

1 Write down all the factors of each number.

    a 15   **b** 20   **c** 32   **d** 35   **e** 60

2 Use your answers to Question 1 to help find the HCF of each pair.

    a 15 and 20   **b** 15 and 60   **c** 20 and 60   **d** 20 and 32

(PS) 3 Mrs Roberts is taking 36 pupils on a visit to a museum. She doesn't want to divide them into groups of unequal size. What possible group sizes could she use?

4 Write down all the common factors for each pair of numbers.

    a 12 and 16   **b** 24 and 36   **c** 10 and 15   **d** 45 and 75

    e 30 and 50   **f** 24 and 40   **g** 15 and 18   **h** 21 and 28

5 Find the HCF of each pair of numbers.

    a 15 and 18   **b** 12 and 32   **c** 12 and 22   **d** 8 and 12

    e 2 and 18   **f** 8 and 18   **g** 18 and 27   **h** 7 and 11

6 Find the largest number that will divide exactly into both numbers in each pair.

    a 20 and 80   **b** 27 and 45   **c** 72 and 140   **d** 140 and 56

    e 75 and 60   **f** 36 and 180   **g** 100 and 150   **h** 120 and 200

7 Simplify each fraction.

    a $\frac{20}{30}$ **b** $\frac{12}{18}$ **c** $\frac{15}{25}$ **d** $\frac{18}{45}$

    e $\frac{24}{64}$ **f** $\frac{28}{35}$ **g** $\frac{108}{240}$ **h** $\frac{72}{81}$

(PS) 8 The teachers of two classes in a primary school needed to put them into groups of equal size. They didn't want to mix the classes. Class A had 36 pupils and class B had 27 pupils.

How many in the largest size groups they can make and how many of these groups would there be?

Investigation: Tests for divisibility

A How can you tell if a number is divisible by 2?

B Write down some numbers that you know are divisible by 3. Make sure they all have more than one digit.

Add up the digits of each of these numbers.

How can you tell if a number is divisible by 3? Write down a rule.

C By looking at the digits of numbers divisible by 4, find a rule for recognising if a number is divisible by 4.

D How can you tell if a number is divisible by 5?

E See how you can combine the rules in parts **A** and **B** to find a rule for recognising if a number is divisible by 6.

F Look at the digits of numbers divisible by 9.

Write down a rule for recognising if a number is divisible by 9.

G Which of these numbers are divisible by:

a 3   **b** 4   **c** 6   **d** 9?

108, 390, 220, 580, 4503, 12 716, 111 111

1.3 Lowest common multiples (LCM)

Learning objective

• To understand and use lowest common multiples

A **multiple** of an integer is the result of multiplying that integer by another integer.

For example, multiplying 3 by 1, 2, 3, 4 and 5 gives 3, 6, 9, 12, 15. So 3, 6, 9, 12, 15 are all multiples of 3.

This also means that any integer that is divisible by 3, giving another integer without a remainder, must be a multiple of 3.

You can find a **common multiple** for any pair of integers by multiplying one by the other. All pairs of integers will have many common multiples, but the **lowest common multiple (LCM)** is generally the most useful. For example, 3 and 4 both have multiples of 12, 24, 36, … but the LCM is 12.

Sometimes the LCM is one of the integers. For example, 12 is a multiple of 12 and of 3, but it is also the LCM of 12 and 3.

• $12 = 12 \times 1$

• $12 = 3 \times 4$

You can use LCMs to help in calculations with fractions that have different denominators, as well as in some real-life problems. You will learn more about this in Chapter 12.

Example 8

Find the lowest common multiple (LCM) of each pair of numbers.

a 3 and 7   **b** 6 and 9

   a Write out the first few multiples of each number.

      3:    3, 6, 9, 12, 15, 18, 21, 24, 27, ...

      7:    7, 14, 21, 28, 35, ...

   You can see that the LCM of 3 and 7 is 21.

   b Write out the multiples of each number.

      6:    6, 12, 18, 24, ...

      9:    9, 18, 27, 36, ...

   You can see that the LCM of 6 and 9 is 18.

Example 9

A baker makes small buns, some of mass 15 g and some of mass 20 g.

He wants to sell them in bags that all have the same mass.

What is the smallest mass he could have in each of these bags?

   The 15 g cakes could be put into batches of mass 15 g, 30 g, 45 g, 60 g, 75 g, ... (all multiples of 15 g).

   The 20 g cakes could be put into batches of mass 20 g, 40 g, 60 g, 80 g, ... (all multiples of 20 g).

   The smallest mass will be the lowest common multiple (LCM) of these numbers, which is 60 g.

Exercise 1C

1 Write down the numbers from the list below that are multiples of:

   a 2   **b** 3   **c** 5   **d** 9.

   10   4   23   18   69   81   8   65   33   72   100

2 Write down the first 10 multiples of each number.

   a 4   **b** 5   **c** 8   **d** 15   **e** 20

3 Use your answers to Question 2 to help find the LCM of the numbers in each pair.

   a 5 and 8   **b** 4 and 20   **c** 6 and 15   **d** 8 and 20

4 Find the LCM of the numbers in each pair.

    a 5 and 9   **b** 5 and 25   **c** 3 and 8   **d** 4 and 6

    e 8 and 12   **f** 12 and 15   **g** 9 and 21   **h** 7 and 11

5 Find the LCM of the numbers in each set.

    a 2, 3 and 5   **b** 3, 4 and 5   **c** 5, 6 and 8   **d** 6, 9 and 15

    e 3, 8 and 12   **f** 5, 12 and 16   **g** 7, 9 and 12   **h** 3, 7 and 11

(PS) 6 In the first year-group of a large school, it is possible to divide the pupils exactly into groups of 24, 30 or 32. Find the smallest number of pupils there could be in this first year-group.

(PS) 7 Two model trains leave the station at the same time and travel around tracks of equal lengths. One completes a circuit in 14 seconds, the other in 16 seconds. How long will it be before they are together again at the station?

(PS) 8 Three friends are walking along a straight pavement.

Suzy has a step size of 14 cm, Kieran has a step size of 15 cm and Miguel has a step size of 18 cm.

They all set off, walking from the same point, next to each other. How far will they have gone before they are all in step with each other again?

(PS) 9 Three men regularly visit their local gym in the evenings.

Joe goes every three days.

John goes every seven days.

James goes every four days.

How many days in a year are they likely all to be in the gym on the same evening?

Challenge: LCM and HCF

A   **a** Two numbers have an LCM of 24 and an HCF of 2. What are they?

    b Two numbers have an LCM of 18 and an HCF of 3. What are they?

    c Two numbers have an LCM of 60 and an HCF of 5. What are they?

B   **a** What are the HCF and the LCM of: **i** 5 and 7   **ii** 3 and 4   **iii** 2 and 11?

    b Two numbers, x and y, have an HCF of 1. What is the LCM of x and y?

C   **a** What are the HCF and LCM of: **i** 5 and 10   **ii** 3 and 18   **iii** 4 and 20?

    b Two numbers, x and y (where y is bigger than x), have an HCF of x.

    What is the LCM of x and y?

1.4 Powers and roots

Learning objective

• To understand and use powers and roots

Squares, cubes and powers

Look at these 3D shapes.

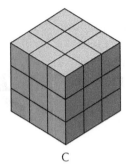

A   B   C

Is shape B twice as big, four times as big or eight times as big as shape A?

How many times bigger is shape C than shape A?

You can calculate the area of each face of shape C by **squaring** the side length.

 3×3 or $3^2 = 9$ You say this as 'three squared'.

You can calculate the volume of the cube by **cubing** the side length.

 $3 \times 3 \times 3$ or $3^3 = 27$ You say this as 'three cubed'.

The small digits 2 and 3 are called **powers**. The power shows how many lots of the number are being multiplied together. A power can be of any size. For example:

• 3^4 is equal to $3 \times 3 \times 3 \times 3 = 9 \times 9 = 81$ and you say it as 'three to the power 4'

• 3^5 is equal to $3 \times 3 \times 3 \times 3 \times 3 = 9 \times 9 \times 3 = 243$ and you say it as 'three to the power 5'.

Example 10

Calculate the value of each number. **a** 4^3 **b** 5.5^2 **c** $(-3)^4$

 a $4^3 = 4 \times 4 \times 4 = 64$

 b $5.5^2 = 5.5 \times 5.5 = 30.25$

  You could also do this on a calculator. Most calculators have a button for squaring, usually marked x^2. They also have a button for cubing, usually marked x^3.

 c $(-3)^4 = -3 \times -3 \times -3 \times -3 = 9 \times 9 = 81$

Square roots and cube roots

To find the area of a square, you 'square' the side.

The inverse process, to work out the side length of a square from its area, is finding the **square root**.

If you know that the area of a square is 25 cm² then the side length will be the square root of 25, written as $\sqrt{25}$. This is the number that will give 25 when you multiply it by itself.

$\sqrt{25} = 5$, because $5 \times 5 = 25$

Square roots can be positive or negative. This is because multiplying two negative numbers gives a positive number.

For example, $4 \times 4 = 16$, but also $-4 \times -4 = 16$.

So $\sqrt{16} = \pm 4$.

If you know that the volume of a cube is 64 cm³, and you want to know its side length, you need to find the **cube root**. This will be the number that, when multiplied by itself and then multiplied by itself again (three 'lots' of the number are multiplied together), gives 64. It is written as $\sqrt[3]{64}$.

$\sqrt[3]{64} = 4$, because $4 \times 4 \times 4 = 64$

Note that you can write the square root simply as $\sqrt{}$, with no small number 2 in front of it, but the cube root must always have a small 3 in front of it, like this: $\sqrt[3]{}$.

Square roots can be positive or negative and a square number is always positive.

A positive cube number can only have a positive cube root and a negative cube number can only have a negative cube root.

$\sqrt[3]{64} = 4$, because $4 \times 4 \times 4 = 64$ but $\sqrt[3]{-64} = -4$, because $-4 \times -4 \times -4 = -64$

Your calculator may have both a square root button, shown as $\boxed{\sqrt{}}$, and a cube root button, shown as $\boxed{\sqrt[3]{}}$.

Practise using your calculator and these buttons by working through the next example, before doing the exercise.

Example 11

Use a calculator to work these out.

a $\sqrt{12.25}$ **b** $\sqrt{33124}$ **c** $\sqrt[3]{2197}$

Depending on your calculator, you might need to select the square root or cube root key before the number. Make sure you know how to use the functions on your calculator.

a $\sqrt{12.25} = 3.5$ **b** $\sqrt{33124} = 182$ **c** $\sqrt[3]{2197} = 13$

Exercise 1D

1 Look back at the diagrams at the beginning of this section. They show cubes made from smaller centimetre cubes. Then copy and complete this table.

Length of side (cm)	1	2	3	4	5	6	7	8	9	10
Area of face (cm²)	1	4	9							
Volume of cube (cm³)	1	8	27							

2 Use your table from Question 1 to work out one value for each number.

  a $\sqrt{4}$ **b** $\sqrt{64}$ **c** $\sqrt{81}$ **d** $\sqrt{100}$ **e** $\sqrt{25}$

  f $\sqrt[3]{27}$ **g** $\sqrt[3]{125}$ **h** $\sqrt[3]{1000}$ **i** $\sqrt[3]{512}$ **j** $\sqrt[3]{729}$

3 Use a calculator to find the value of each number.

  a 13^2 **b** 13^3 **c** 15^2 **d** 15^3 **e** 21^2 **f** 21^3

  g 1.4^2 **h** 1.8^3 **i** 2.3^3 **j** 4.5^2 **k** 12^3 **l** 1.5^3

4 Use a calculator to find the value of each number.

  a 2^4 **b** 3^5 **c** 3^4 **d** 2^5 **e** 4^4 **f** 5^4

  g 7^4 **h** 8^3 **i** 2^7 **j** 2^9 **k** 2^{10} **l** 3^{10}

5 Without using a calculator, write down the value of each number.

> **Hint** Use your table from Question 1 to help you.

  a 20^2 **b** 30^3 **c** 50^3 **d** 20^5 **e** 70^2 **f** 200^3

6 $10^2 = 100$, $10^3 = 1000$

Copy and complete this table.

Number	100	1000	10 000	100 000	1 000 000	10 000 000
Power of 10	10^2	10^3				

7 Find two values of x that make the equation true, each time.

  a $x^2 = 36$ **b** $x^2 = 121$ **c** $x^2 = 144$ **d** $x^2 = 2.25$

  e $x^2 = 196$ **f** $x^2 = 5.76$ **g** $x^2 = 2.56$ **h** $x^2 = 3600$

8 **a** Work out the value of each number.

    i 1^2 **ii** 1^3 **iii** 1^4 **iv** 1^5 **v** 1^6

  b Write down the value of 1^{123}.

9 **a** Work out the value of each number.

    i $(-1)^2$ **ii** $(-1)^3$ **iii** $(-1)^4$ **iv** $(-1)^5$ **v** $(-1)^6$

  b Write down the value of: **i** $(-1)^{223}$ **ii** $(-1)^{224}$.

(PS) 10 You should see from your table in Question 1 that 64 is a square number (8^2) and a cube number (4^3).

  a One other cube number (apart from 1) in the table is also a square number. Which is it?

> **Hint** Look at the pattern of cube numbers so far, for example, 1^3, 4^3, ...

  b Which is the next cube number that is also a square number?

How many squares are there on a chessboard?

Hint A computer spreadsheet is useful for this activity.

The answer is not 64!

For example, in this square there are actually five squares.

There are four this size and one this size .

In this square there are actually 14 squares.

There are nine this size , four this size and one this size .

By drawing increasingly larger 'chessboards', work out how many squares there are and see if you can spot the pattern.

1.5 Prime factors

Learning objectives

- To understand what prime numbers are
- To find the prime factors of an integer

Key words	
factor tree	index form
prime factor	prime number

You may remember that a **prime number** is an integer that has only two factors, itself and one.

These are the first ten prime numbers.

2, 3, 5, 7, 11, 13, 17, 19, 23, 29

A **prime factor** of an integer is a factor that is also a prime number.

Therefore the prime factors of an integer are the prime numbers that will multiply together to give that integer.

For example:

- 6 can be written as a product of its prime factors, as 2×3
- 12 can be written as a product of its prime factors, as $2 \times 2 \times 3$
- 16 can be written as a product of its prime factors, as $2 \times 2 \times 2 \times 2$.

From this, you can see that some numbers have repeated prime factors.

One useful way to find the prime factors of any integer is to draw up a **factor tree**.

Find the prime factors of 18.

Using a prime factor tree, start by splitting 18 into 3×6.

Then split the 6 into 3×2.

You could also split 18 into 2×9 and then split the 9 into 3×3.

Keep splitting the factors up until you only have prime numbers at the ends of the 'branches'.

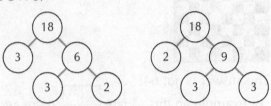

Whichever pair of factors you start with, you will always finish with the same set of prime factors.

So, $18 = 2 \times 3 \times 3$

$\qquad = 2 \times 3^2$

When the number is written as 2×3^2, it is in **index form**.

There is another way to calculate the prime factors of a number.

- Start with the smallest prime number that is a factor of the number.
- Divide that prime number into the integer as many times as possible.
- Then try the next smallest prime number that is a factor of the number.
- Carry on until you reach 1.

Find the prime factors of 24.

Use the division method.

$$
\begin{array}{r|r}
2 & 24 \\
\hline
2 & 12 \\
\hline
2 & 6 \\
\hline
3 & 3 \\
\hline
 & 1
\end{array}
$$

So, $24 = 2 \times 2 \times 2 \times 3 = 2^3 \times 3$.

Exercise 1E

1. These numbers are written as products of their prime factors. What are the numbers?

   a $2 \times 2 \times 3$ b $2 \times 3 \times 3 \times 5$ c $2 \times 2 \times 3^2$ d $2 \times 3^3 \times 5$ e $2 \times 3 \times 5^2$

2. Use a prime factor tree to work out the prime factors of each number.

   a 8   b 10   c 16   d 20   e 28

   f 34   g 35   h 52   i 60   j 180

3. Use the division method to work out the prime factors of each number.

   a 42   b 75   c 140   d 250   e 480

4 Find the prime factors of all the numbers from 2 to 20.

5 **a** Which numbers in Question 4 only have one prime factor?
b What special name is given to these numbers?

6 100 can be written as a product of its prime factors as $100 = 2 \times 2 \times 5 \times 5 = 2^2 \times 5^2$.
a Write down 200 as a product of its prime factors, in index form.
b Write down 50 as a product of its prime factors, in index form.
c Write down 1000 as a product of its prime factors, in index form.
d Write down the prime factors of one million, in index form.

7 The smallest number with exactly two *different* prime factors is $2 \times 3 = 6$.
a What is the next smallest number with exactly two *different* prime factors?
b What is the smallest number with exactly three *different* prime factors?

8 **a** What are the prime factors of 32? Give your answer in index form.
b Write down the prime factors of 64, in index form.
c Write down the prime factors of 128, in index form.
d Write down the prime factors of 1024, in index form.

9 **a** What are the prime factors of 60?
b What are the prime factors of 100?
c Use your answers to parts **a** and **b** to find the LCM of 60 and 100.

d Explain how you found your answer in part c.

10 Find the LCM of 84 and 140.

Challenge: LCM and HCF in Venn diagrams

A Use the diagrams to work out the HCF and LCM of each pair of numbers.

a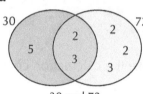
30 and 72

b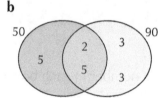
50 and 90

c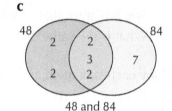
48 and 84

B The prime factors of 120 (including repeats) are 2, 2, 2, 3 and 5.
The prime factors of 150 are 2, 3, 5 and 5.
Put these numbers into a diagram like those above.
Use your diagram to work out the HCF and LCM of 120 and 150.

C The prime factors of 210 are 2, 3, 5 and 7.
The prime factors of 90 are 2, 3, 3 and 5.
Put these numbers into a diagram like those above.
Use your diagram to work out the HCF and LCM of 210 and 90.

Ready to progress?

I can find square and cube numbers and square and cube roots.
I can use a calculator to work out powers of numbers.
I can find common factors for pairs of numbers.

I can multiply and divide negative numbers, for example, $-5 \times 3 = -15$.
I know that the square roots of positive numbers can have two values, one positive and one negative.
I can find the lowest common multiple (LCM) for pairs of numbers.
I can find the highest common factor (HCF) for pairs of numbers.
I can write a number as the product of its prime factors.

Review questions

1   Which of these statements are true?

   a   All odd numbers are prime numbers.

   b   All even numbers are multiples of two.

   c   All prime numbers are odd numbers.

   d   All prime numbers between 10 and 100 are odd numbers.

   e   The integer 2 is the only even prime number.

   f   There are five prime numbers less than 10.

2   a   Put these values in order of size, with the smallest first.

      5^2 3^3 2^4

   b   You are given that $4^5 = 1024$.
      What is the value of 4^7?

3   Copy each statement and write a number in each box to make the calculation correct.

   a $\square \times \square = -10$ b $\square \div \square = -10$

4   a   Two numbers multiplied together give −28. They add together to give 3.
      What are the two numbers?

   b   Two numbers multiplied together give −28, but added together they give −3.
      What are the two numbers?

   c   The squares of two numbers add up to 61.
      What are the two numbers?

5   A room measures 450 cm by 350 cm. Find the side of the largest square tile that could be used to tile the whole floor, without any cutting.

6 a Find the shortest length of fabric that can be divided exactly into equal parts of length 9 m, 12 m, 36 m or 48 m.

   b Find the longest length that can be divided exactly into any of these lengths.

 7 In a furniture shop Frank saw this footstool.

The SQUARE footstool
Base 36 cm, width 36 cm, height 49 cm
£256

Frank noticed that all of the details were square numbers.

He wondered: 'If this were a medium square footstool, what would be the size and price of the next-smallest square footstool and the price and size of the next-largest square footstool?'

   a Find the cost and the size of the next-smallest square footstool.

   b Find the cost and the size of the next-largest square footstool.

 8 Rafael was away for the night. He was trying to sleep, but there was too much noise around him.

   • His travel clock ticked every 2 seconds.

   • A tap was dripping every 5 seconds.

   • Someone in the next room snored every 6 seconds.

   • A car seemed to go past outside every 7 seconds.

   He noticed that all four things happened together on the stroke of midnight.

   a After how many more seconds will all four things happen again?

   b How many times will all four things happen together again between midnight and one o'clock?

9 Copy this Venn diagram.

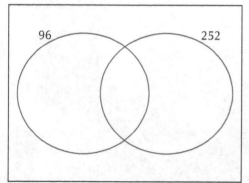

96   252

   a Write the prime factors of 96 and 252 into the correct parts of your Venn diagram.

   b Use your diagram to find:

      i   the HCF of 96 and 252   ii   the LCM of 96 and 252.

Challenge
Blackpool Tower

Blackpool Tower is a tourist attraction in Blackpool, Lancashire, England. It was inspired by the Eiffel Tower in Paris. It opened to the public on 14 May 1894. It is 518 feet 9 inches tall and now attracts about 500 000 visitors a year.

Inside the tower there is a circus, an aquarium, a ballroom, restaurants, a children's play area and amusements.

1 The largest tank in the aquarium holds 32 000 litres of water.
   There are approximately 4.5 litres to a gallon.
   How many gallons of water does the tank hold?

2 When the circus in the base of the tower first opened to the public, the admission fee was 6 old pence. Before Britain introduced decimal currency in 1971 there were 240 old pence in a pound.

   a What fraction, in its simplest form, is 6 old pence of 240 old pence?

   b What is the equivalent value of 6 old pence in new pence?

3 a In January 2008, it cost €12 to visit the Eiffel Tower and £9.50 to visit Blackpool Tower. The exchange rate in January 2008 was £1 = €1.35. Which tower was cheaper to visit? By how much? Give your answer in pounds and pence.

b The Eiffel Tower is about 325 m high. Blackpool Tower is about 519 feet high. 1 m ≈ 3.3 feet. How many times taller is the Eiffel Tower than the Blackpool Tower?

c The Eiffel Tower gets 6.7 million visitors a year. How many times more popular is it than the Blackpool Tower with tourists?

4 The Ballroom floor measures 36.58 m by 36.58 m. It comprises 30 602 separate blocks of mahogany, oak and walnut. Assuming that every block is equal in area, what is the area, in square centimetres, of each block?

Give your answer to the nearest square centimetre.

5 When it is lit up, the tower has 10 000 light bulbs, using an average of 15 watts per hour each. The cost of electricity is 12p per kilowatt hour (1 kWh = 1000 watts per hour). Calculate the approximate yearly electricity bill for the lights, assuming they are lit for 12 hours per day.

6 The circus ring, when flooded, can hold up to 190 000 litres of water to a depth of 140 cm. (1 litre = 1000 cm³)

a How many cubic centimetres is 190 000 litres?

b Assuming that the circus ring is circular, the formula for working out the radius, given the volume, V, and the depth, d, is:

$$r = \sqrt{\frac{V}{\pi \times d}}$$

Work out the radius of the circus ring. Give your answer in metres.

7 An approximate formula for how far you can see, D kilometres, when you are m metres above the ground is:

$D = \sqrt{13m}$

The coast of the Isle of Man is 42 km from Blackpool. Can you see it from the observation deck of the tower, which is 120 m above the ground?

Show your working clearly.

2

Geometry

This chapter is going to show you:

- how to identify alternate and corresponding angles
- how to classify quadrilaterals according to their geometrical properties
- how to rotate a shape
- how to translate a shape
- how to construct perpendicular lines and angle bisectors.

You should already know:

- how to identify parallel and perpendicular lines
- how to calculate angles at a point, angles on a line and opposite angles
- how to calculate angles in a triangle
- how to recognise the symmetry of quadrilaterals
- how to plot coordinates in all four quadrants
- how to use a pair of compasses to draw a circle.

About this chapter

Architects use the properties of angles and shapes to design buildings. When you look at a building you can often see triangles, quadrilaterals and other shapes.

Understanding angles, lines and shapes and being able to construct them accurately are important in all types of design. For example, many fabrics and wallpapers use angles, lines and shapes that have been rotated, reflected and repeated to make a pattern that looks good and is also easy to reproduce. When fabrics and wallpaper are printed, the design is reproduced by translating it repeatedly.

This chapter will show you some of the properties of angles and shapes.

2.1 Angles in parallel lines

Learning objective

* To calculate angles in parallel lines

Key words

allied angles

alternate angles

corresponding angles

transversal

A line that intersects a set of parallel lines is called a **transversal**.

Notice in the diagram that eight distinct angles are formed by a transversal.

The two angles marked on this diagram are equal and are called **corresponding angles**.

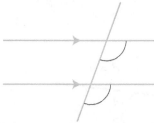

To help you remember which angles are corresponding angles, notice that they occur where the lines make the shape of a letter F.

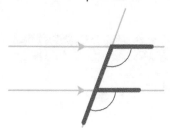

The two angles marked on this diagram are equal and are called **alternate angles**.

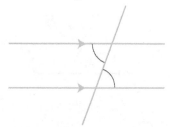

Look for places where the lines make the shape of the letter Z to identify alternate angles.

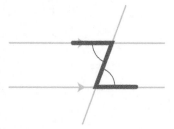

Example 1

Look at this diagram.

a Name pairs of angles that are alternate angles.

b Name pairs of angles that are corresponding angles.

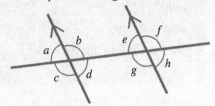

a There are two pairs of alternate angles.

$b = g$

$d = e$

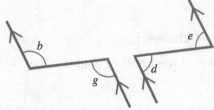

b There are four pairs of corresponding angles.

$a = e$ $b = f$ $c = g$ $d = h$

Example 2

Calculate the sizes of the angles labelled p, q and r.

Give reasons for your answers.

a   **b**   **c**

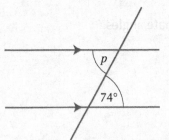 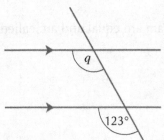 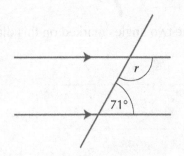

a $p = 74°$ Alternate angles are equal.

b $q = 123°$ Corresponding angles are equal.

c The angle adjacent to r is 71°. Alternate angles are equal.

So $r = 109°$.   Angles on a line add up to 180°.

Exercise 2A

1 Copy and complete each sentence.

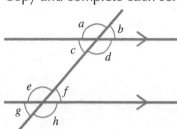

a *a* and … are corresponding angles.   **b** *b* and … are corresponding angles.

c *c* and … are corresponding angles.   **d** *d* and … are corresponding angles.

e *e* and … are alternate angles.   **f** *f* and … are alternate angles.

2 Work out the size of each lettered angle in these diagrams.

a

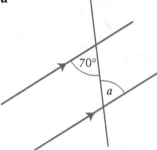

b

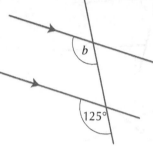

c

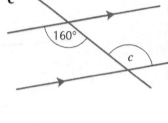

d

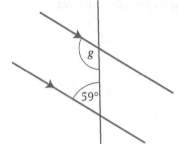

e

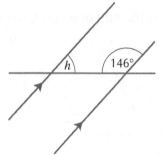

MR **3** Which diagram is the odd one out? Give a reason for your answer.

a 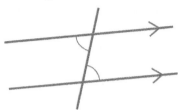   **b**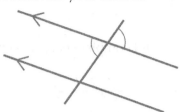

c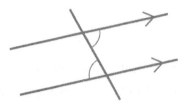

MR **4** Which diagram is the odd one out? Give a reason for your answer.

a    **b**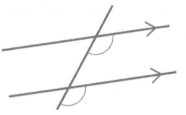

c

MR **5** Which diagram is the odd one out? Give a reason for your answer.

a 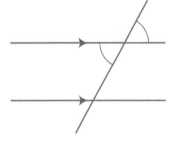   **b**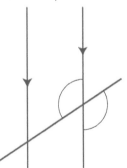

c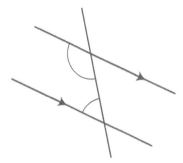

6 Calculate the size of each unknown angle.

State whether it is an alternate angle or a corresponding angle.

a
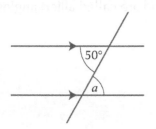

b
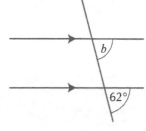

c
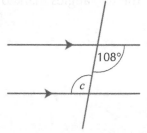

d
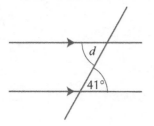

e
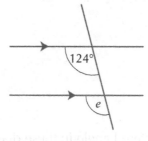

f
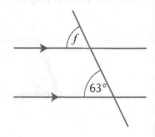

(MR) **7** Calculate the size of each unknown angle.

Explain how you worked out your answers.

a
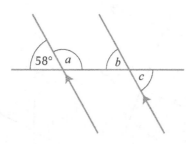

b
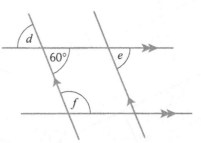

8 Look at this a parallelogram.

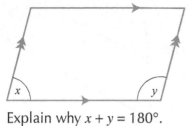

> **Hint** Extend the base of the parallelogram.

Explain why $x + y = 180°$.

(MR) **9** Calculate the value of f on the diagram.

Draw a diagram to explain your answer.

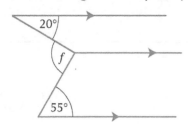

Mathematical reasoning: More about parallel lines

The two angles marked on this diagram add up to 180° and are called **allied angles**.

Look for where the lines make the shape of the letter C to identify allied angles.

Calculate the size of each lettered angle in these diagrams.

Give reasons for your answers.

A

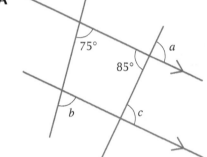

B

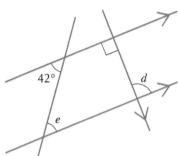

C

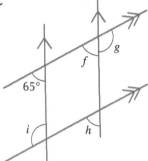

D

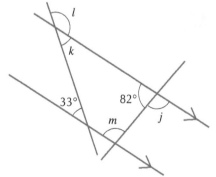

E

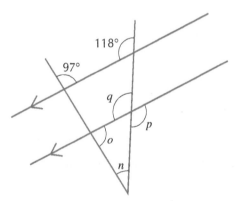

2.2 The geometric properties of quadrilaterals

Learning objective

- To know the geometric properties of quadrilaterals

Key word

geometric properties

Read these descriptions about some quadrilaterals carefully and learn all their **geometric properties**.

Square

- It has four equal sides.
- It has four right angles.
- Its opposite sides are parallel.
- Its diagonals bisect each other at right angles.
- It has four lines of symmetry.
- It has rotational symmetry of order four.

Rectangle

- It has two pairs of equal sides.
- It has four right angles.
- Its opposite sides are parallel.
- Its diagonals bisect each other.
- It has two lines of symmetry.
- It has rotational symmetry of order two.

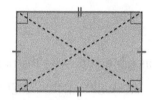

Parallelogram

- It has two pairs of equal sides.
- It has two pairs of equal angles.
- Its opposite sides are parallel.
- Its diagonals bisect each other.
- It has no lines of symmetry.
- It has rotational symmetry of order two.

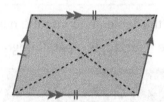

Rhombus

- It has four equal sides.
- It has two pairs of equal angles.
- Its opposite sides are parallel.
- Its diagonals bisect each other at right angles.
- It has two lines of symmetry.
- It has rotational symmetry of order two.

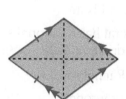

Kite

- It has two pairs of adjacent sides of equal length.
- It has one pair of equal angles.
- Its diagonals intersect at right angles.
- It has one line of symmetry.

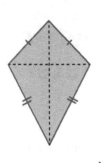

Arrowhead

- It has two pairs of adjacent sides of equal length.
- It has one pair of equal angles.
- Its diagonals intersect at right angles outside the shape.
- It has one line of symmetry.

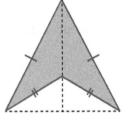

Trapezium

- It has one pair of parallel sides.
- Some trapezia have one line of symmetry.

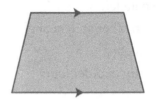

Exercise 2B

1 Copy this table and write the name of each quadrilateral in the correct column.

square   rectangle   parallelogram   rhombus   kite   arrowhead   trapezium

No lines of symmetry	One line of symmetry	Two lines of symmetry	Four lines of symmetry

2 Copy this table and write the name of each quadrilateral in the correct column.

square   rectangle   parallelogram   rhombus   kite   arrowhead   trapezium

Rotational symmetry of order one	Rotational symmetry of order two	Rotational symmetry of order four

3 A quadrilateral has four right angles and rotational symmetry of order two. What type of quadrilateral is it?

4 A quadrilateral has rotational symmetry of order two and no lines of symmetry. What type of quadrilateral is it?

MR **5** Read what Rachel says.

Is she right or wrong?

Explain your answer.

A quadrilateral with four equal sides must be a square.

6 Read what Robert says.

Is he right or wrong?

Explain your answer.

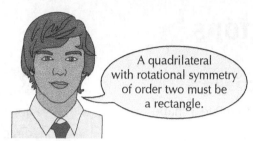

A quadrilateral with rotational symmetry of order two must be a rectangle.

7 Priya knows that a square is a special kind of rectangle because it is a rectangle with four equal sides.

Write down the names of other quadrilaterals that could also be used to describe a square.

8 Look at the kite ABCD.

Use symmetry to explain why $p = q$.

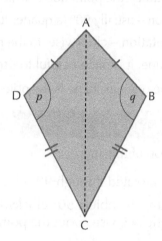

Investigation: Rectangles into squares

The 3 by 2 rectangle below is to be cut into squares along its grid lines.

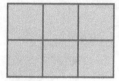

You can do this in two different ways.

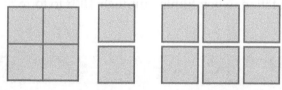

Three squares   Six squares

Use squared paper to show the number of ways that rectangles of different sizes can be cut into squares.

2.3 Rotations

Learning objective

• To understand how to rotate a shape

Key words

| angle of rotation | centre of rotation |
| direction of rotation | rotation |

You have seen how a 2D shape can be rotated.

To describe the **rotation** of a 2D shape accurately, you need to know three facts. These are:

- the **centre of rotation** – the point about which the shape rotates
- the **angle of rotation** – usually 90° (a quarter-turn), 180° (a half-turn) or 270° (a three-quarter turn)
- the **direction of rotation** – clockwise (to the right) or anticlockwise (to the left).

When you rotate a shape, it is often helpful to use tracing paper.

As with reflections, the original shape is the object and the rotated shape is the image.

Example 3

Describe the rotation of this flag.

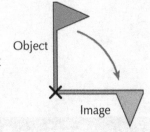

The flag has been rotated through 90° clockwise about the point X.

Notice that rotating the object 90° clockwise is the same as rotating it through 270° anticlockwise about the point X.

Example 4

Describe the rotation of this triangle.

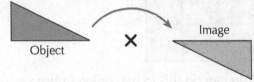

The right-angled triangle has been rotated through 180° clockwise about the point X.

Notice that the object can be rotated either clockwise or anticlockwise about the centre of rotation, when turning through 180° to form the image.

Example 5

Describe the rotation of triangle ABC.

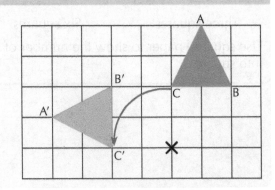

Triangle ABC has been rotated through 90° anticlockwise, about the point X, onto triangle A′B′C′.

Describe the rotation of triangle ABC.

Triangle A'B'C' is the image of triangle ABC after a rotation of 90° anticlockwise about the origin O(0, 0).

The coordinates of the vertices of the object are A(1, 2), B(4, 4) and C(4, 2).

The coordinates of the vertices of the image are A'(−2, 1), B'(−4, 4) and C'(−2, 4).

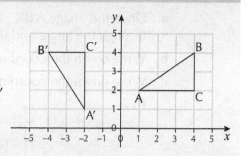

Exercise 2C

1. Copy each flag below and draw its image after it has been rotated about the point marked X, through the angle indicated. Use tracing paper to help.

   a   b   c

   90° clockwise   90° anticlockwise   180° clockwise

2. Copy each shape onto a square grid.

   Then draw its image after it has been rotated, about the point marked X, through the angle indicated. Use tracing paper to help.

   a 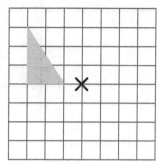   b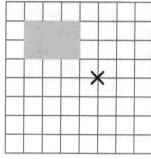

   180° clockwise   90° anticlockwise

   c 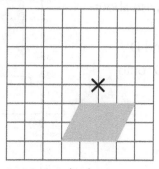   d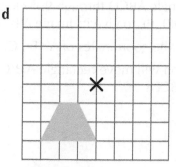

   180° anticlockwise   270° clockwise

3 Copy each right-angled triangle ABC onto a coordinate grid, with axes for *x* and *y* both numbered from −5 to 5.

   a Draw the image A′B′C′ after triangle ABC has been rotated about the origin O, through the angle and direction indicated.

   b Write down the coordinates of the object.

   c Write down the coordinates of the image.

i

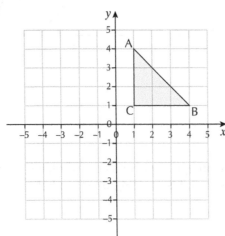

90° anticlockwise

ii

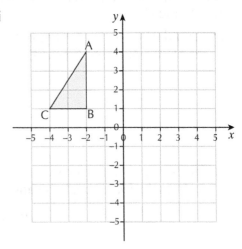

90° clockwise

iii

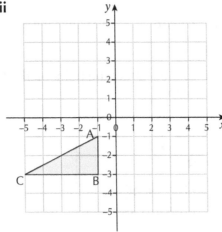

180° anticlockwise

iv

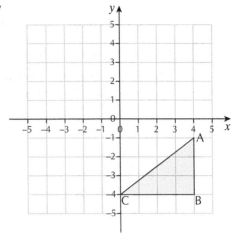

180° clockwise

4 Copy this rectangle onto a coordinate grid.

   a Rotate rectangle ABCD through 90° clockwise about the origin O(0, 0), to give its image A′B′C′D′.

   b Write down the coordinates of A′, B′, C′ and D′.

   c What rotation will move rectangle A′B′C′D′ onto rectangle ABCD?

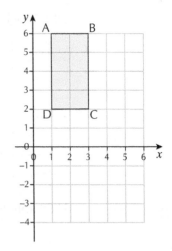

5 Copy parallelogram A onto a coordinate grid, with axes for *x* and *y* both numbered from −5 to 5.

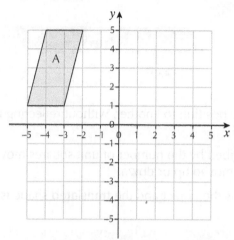

a Rotate parallelogram A through 90° anticlockwise about the point (2, 1). Label the image B.

b Write down two different rotations that will move parallelogram B onto parallelogram A.

Challenge: Finding the centre of rotation

The grid shows three identical isosceles triangles.

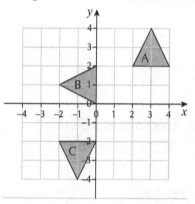

A Describe the rotation that moves A onto B.

B Describe the rotation that moves A onto C.

C Describe the rotation that moves C onto B.

 The centre of rotation is different for each one.

2.4 Translations

Learning objective

- To understand how to translate a shape

Key words

| translate | translation |

A **translation** is a movement of a 2D shape from one position to another, without reflecting it or rotating it.

The distance and direction of the translation is described by the number of unit squares moved to the right or left, followed by the number of unit squares moved up or down.

As with reflections and rotations, the original shape is the object and the **translated** shape is the image.

Example 7

Translate triangle A onto triangle B by the translation 3 units right and 2 units up.

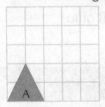

Points on triangle A are translated onto triangle B, as shown by the arrows.

When an object is translated onto its image, every point on the object moves the same distance, in the same direction.

Example 8

Describe the translation of rectangle ABCD.

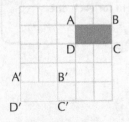

The rectangle ABCD has translated onto rectangle A'B'C'D' by the translation 3 units left and 3 units down.

Exercise 2D

1 Describe the translation:

   a from A to B   **b** from A to C

   c from A to D   **d** from A to E

   e from B to D   **f** from C to E

   g from D to E   **h** from E to A.

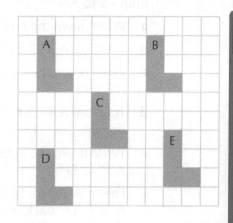

2 Copy the triangle ABC onto squared paper, with axes for x and y both numbered from −5 to 5. Label the triangle P.

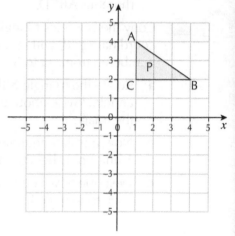

   a Write down the coordinates of the vertices of triangle P.

   b Translate triangle P 6 units left and 2 units down. Label the new triangle Q.

   c Write down the coordinates of the vertices of triangle Q.

   d Translate triangle Q 5 units right and 4 units down. Label the new triangle R.

   e Write down the coordinates of the vertices of triangle R.

   f Describe the translation that translates triangle R onto triangle P.

 3 Copy the trapezium ABCD onto a coordinate grid, with axes for x and y both numbered from −5 to 5.

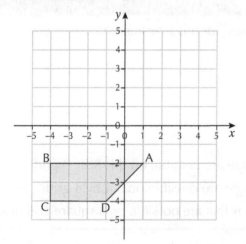

   a Translate the trapezium 4 units right and 5 units up.

   b Write down the coordinates of the vertices of the image.

   c Describe what properties of the trapezium have changed and what have stayed the same after this translation.

4 Copy the rhombus ABCD onto a coordinate grid, with axes for x and y both numbered from −5 to 5.

  a Write down the coordinates of ABCD.

  b Translate ABCD 2 units left and 6 units up.

  c Write down the coordinates of the image A′B′C′D′.

  d Translate A′B′C′D′ 3 units left and 4 units down.

  e Write down the coordinates of the image A″B″C″D″.

  f Describe the translation that translates A″B″C″D″ back onto the original rhombus ABCD.

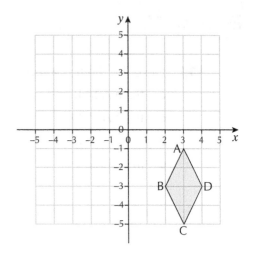

(MR) **5** Copy the isosceles triangle X onto a coordinate grid, with axes for x and y both numbered from −5 to 5.

  a Rotate the triangle X through 90° anticlockwise about the origin O. Label the image Y.

  b Now translate triangle Y 4 units right and 1 unit up. Label the image Z.

  c Describe how you would move triangle Z back to triangle X.

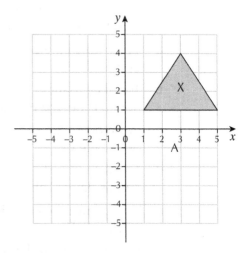

Investigation: Dotty translations

Use centimetre-squared dotty paper or a pin-board for this investigation.

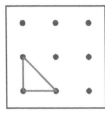

A How many different translations of the triangle are possible on this 3 by 3 grid?

B How many different translations of this triangle are possible on a 4 by 4 grid?

C Investigate the number of different translations that are possible on a square grid of any size.

2.5 Constructions

Learning objectives

- To construct the mid-point and the perpendicular bisector of a line
- To construct an angle bisector

Key words

angle bisector	bisect
construction	perpendicular bisector

The next two examples will show you two important geometric **constructions**. They are useful because they give exact measurements and are therefore used by architects and in design and technology. You will need a sharp, hard pencil, a ruler, compasses and a protractor.

Always leave all of your construction lines on the diagrams.

- The first construction enables you to find the mid-point of one line and construct a second line at right angles to it at that point. This line is called the **perpendicular bisector** of the line.
- The second construction produces a line that is the bisector of an angle.

Perpendicular means 'at right angles to' and bisector means it **bisects** it, or cuts it in half.

Example 9

Construct the mid-point and the perpendicular bisector of the line AB.

- Draw a line segment AB of any length.

  A ————————————— B

- Set your compasses to a radius greater than half the length of AB.
- Draw two arcs, with the centre at A, one above and one below AB.
- With your compasses still set at the same radius, draw two arcs with the centre at B, to intersect the first two arcs at C and D.
- Join C and D to intersect AB at X. X is the midpoint of the line AB.

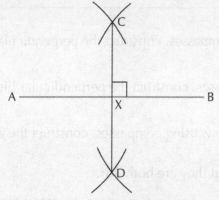

- The line CD is the perpendicular bisector of the line AB.

The second construction enables you to bisect an angle accurately and draw the **angle bisector**.

Example 10

Construct the angle bisector of the angle ABC.

- Draw an angle ABC of any size.

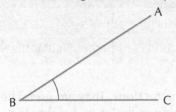

- Set your compasses to any radius and, with the centre at B, draw an arc to intersect BC at X and AB at Y.
- With your compasses set to any radius, draw two arcs with the centres at X and Y, to intersect at Z.
- Join BZ.

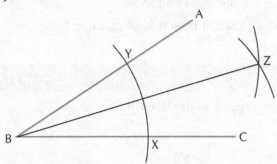

- BZ is the bisector of the angle ABC.
- Then ∠ABZ = ∠CBZ.

Exercise 2E

1. Draw a line AB that is 10 cm long. Using compasses, construct the perpendicular bisector of the line.

2. Draw a line CD of any length. Using compasses, construct the perpendicular bisector of the line.

3. Use a protractor to draw an angle of 80°. Now, using compasses, construct the angle bisector of this angle.

   Measure the two angles formed, to check that they are both 40°.

4. Use a protractor to draw an angle of 140°. Now, using compasses, construct the angle bisector of this angle. Measure the two angles formed, to check that they are both 70°.

5. Draw a line AB that is 8 cm long.

   a Construct the perpendicular bisector of AB.

   b By measuring the length of the perpendicular bisector, draw a rhombus with diagonals of length 8 cm and 5 cm.

6 Draw a circle of radius 6 cm. Label the centre O.
Draw a line AB of any length across the circle, as in the diagram.

AB is a chord.

Construct the perpendicular bisector of AB. Extend the
perpendicular, if necessary, to make a diameter of the circle.

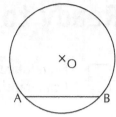

MR **7** Construct an angle of 60°.

First, draw a line AB of any length. Set your compasses to a
radius of about 4 cm. With centre at A, draw a large arc to
intersect the line at X. Using the same radius, and with the
centre at X, draw an arc to intersect the first arc at Y.
Join A and Y: ∠YAX is 60°.

Explain how you could use this construction to construct
angles of 30° and 15°.

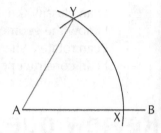

Activity: Construct a line parallel to a given line and passing through a given point

A **a** Copy the diagram, which shows a line AB with a point C above it.

C
X

A————————B

b Set your compasses to a suitable radius.

Draw an arc with centre C to intersect AB at P.

c Keep your compasses set to the same radius.

Draw an arc with centre P to intersect AB at Q.

d Still using the same radius, draw arcs from
C and Q to intersect at R.

e Join CR.

CR is parallel to AB.

B Copy each diagram.

Construct a line parallel to the given line and passing through the given point.

a

b

Ready to progress?

I can rotate a simple shape about a point.
I can translate a shape.

I know the angle properties of parallel lines.
I can classify the different types of quadrilateral.
I know the geometric properties of quadrilaterals.
I can rotate a shape about a given centre of rotation on a coordinate grid.
I can construct perpendicular lines and bisect angles.

Review questions

1 Copy each of these right-angled triangles and draw the image after it has been rotated about the point marked X through the angle indicated. Use tracing paper to help.

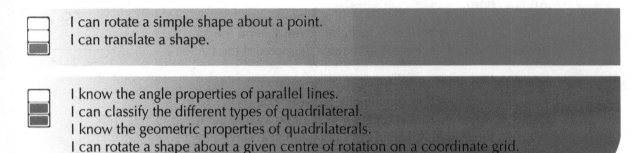

a
90° clockwise

b
180° clockwise

c
90° anticlockwise

(MR) 2 Copy the square P onto a square grid.

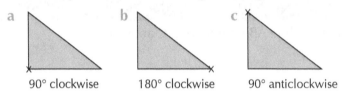

a Translate P 3 units right and 4 units down. Label the image Q.

b Translate Q 2 units left and 1 unit up. Label the image R.

c What translation moves R back to P?

3 Calculate the size of each unknown angle.

Give reasons for your answers.

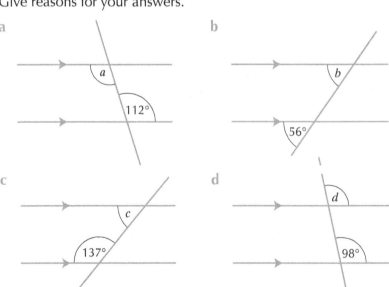

a

b

c

d

4 The diagram shows two isosceles triangles drawn inside a parallelogram.

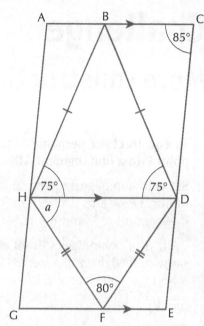

a Write down two other angles that are 75°.

Give reasons for your answer.

b Calculate the size of the angle DHF, marked *a* on the diagram.

Give a reason for your answer.

c Angle BCD is 85°. Calculate the size of angle BAH.

Give a reason for your answer.

5 This 2D shape is made from three identical white rhombi and three identical grey rhombi.

The sides of each rhombus are all the same length.

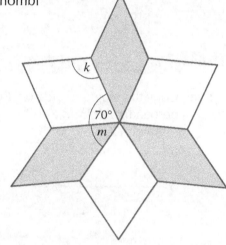

a Calculate the size of the angle labelled *k*.

Give a reason for your answer.

b Calculate the size of the angle labelled *m*.

Show your working.

6 Copy the right-angled triangle ABC onto a coordinate grid, with axes for *x* and *y* both numbered from −5 to 5.

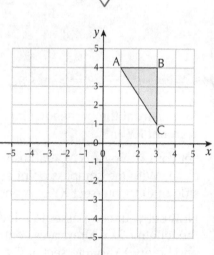

a Write down the coordinates of the vertices of triangle ABC.

b Rotate the triangle through 180° clockwise about the origin O. Label the image A′B′C′.

c Write down the coordinates of the vertices of triangle A′B′C′.

d What do you notice about the coordinates of the vertices of triangle A′B′C?

e The coordinates of the vertices of another triangle PQR are P(2, 1), Q(3, 5), R(4, 1).

Triangle PQR is rotated through 180° clockwise about the origin O to give triangle P′Q′R′.

Write down the coordinates of the vertices of triangle P′Q′R′.

Challenge

More constructions

To construct the perpendicular from a point P to a line segment AB

Set your compasses to any suitable radius. Draw arcs centred on P to intersect AB at X and Y.

With your compasses still set at the same radius, draw arcs centred on X and Y to intersect at Z below AB.

Join PZ.

PZ is the perpendicular from P to AB.

1 a Copy the diagram and construct the perpendicular from X to the line segment YZ.

Y ——————————— Z

X×

b Copy the diagram and construct the perpendicular from C to the line segment DE.

C×

To construct the perpendicular from a point Q on a line segment XY

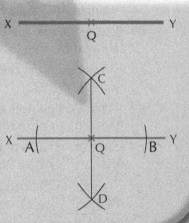

Set your compasses to a radius that is less than half the length of XY. With the compass point at Q, draw two arcs on either side of Q to intersect XY at A and B.

Now set your compasses to a radius that is greater than half the length of XY. Draw arcs centred on A, and then B, above and below XY to intersect at C and D.

Join CD.

CD is the perpendicular to XY from the point Q.

2 a Copy the diagram and construct the perpendicular from the point A on the line segment BC.

b Construct the right-angled triangle XYZ accurately.

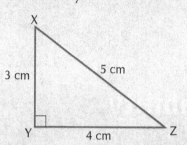

To construct the inscribed circle of a triangle

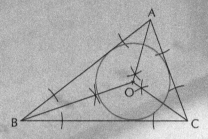

Draw a triangle ABC with sides of any length.

Construct the angle bisectors for each of the three angles.

The three angle bisectors will meet at a point O in the centre of the triangle. Using O as the centre, draw a circle to touch all three sides of the triangle.

The circle is the inscribed circle of the triangle.

To construct the circumscribed circle of a triangle

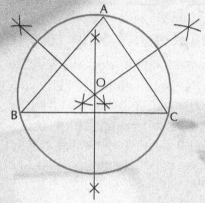

Draw a triangle ABC with sides of any length.

Construct the perpendicular bisector for each of the three sides. The three perpendicular bisectors will meet at a point O. Using O as the centre, draw a circle to touch the three vertices of the triangle.

The circle is the circumcircle of the triangle, and O is the circumcentre.

3 Draw the triangle XYZ and construct the inscribed circle.

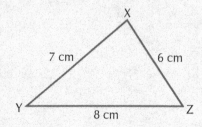

4 Draw the triangle XYZ and construct the circumscribed circle.

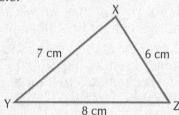

3

Probability

This chapter is going to show you:

- how to work with a probability scale
- how to recognise mutually exclusive and non-exclusive outcomes and events
- how to work out probabilities, using sample spaces and Venn diagrams where necessary
- how to use experimental probability to make predictions.

You should already know:

- what chance and probability are
- how to collect data from a simple experiment
- how to record data in a table or chart.

About this chapter

What is the probability that you will ever travel in space?

One hundred years ago, the chance of this was nil, that is, it was impossible, but now the chance is increasing every decade. Scientists predict that many pupils in schools now will have a fair chance of travelling into space one day in their lifetime. They calculate the probabilities by working out what is technically possible, and who might be able to afford it.

We do not know for certain if mass space travel will happen but, by studying probability, we can understand how likely it is to happen and how the scientists work it out.

3.1 Probability scales

Learning objective

- To use a probability scale to represent a chance

Key words

equally likely	event
outcome	probability
probability scale	

When you do something such as rolling a dice, this is called an **event**.

The possible results of the event are called its **outcomes**. For example, rolling a dice has six possible outcomes: scoring 1, 2, 3, 4, 5 or 6.

You can use **probability** to decide how likely it is that different outcomes will happen.

Equally likely outcomes

Equally likely outcomes are those that all have the same chance of happening. For example, when you roll a dice, there are six different possible outcomes. This is because it could land so that any one of its six numbers shows on top.

The probability of an equally likely outcome is:

$$P(outcome) = \frac{\text{the number of ways the outcome could occur}}{\text{the total number of possible outcomes}}$$

There is only one way a normal dice can show a 6 when it lands, so only one of its possible outcomes will be the one you want.

There are six numbers on the faces, so there are six possible outcomes. Therefore:

$$P(6) = \frac{1}{6}$$

If you throw two dice and add their scores there are several ways that you could score 6, because there is a much larger number of possible outcomes.

Example 1

What is the probability of scoring a number less than 5 when you roll a dice?

There are four possible outcomes that give you a number less than 5: 1, 2, 3 and 4.

There are six different possible outcomes altogether, when you roll a dice: 1, 2, 3, 4, 5 and 6.

So P(rolling a dice and getting a number less than 5) is $\frac{4}{6} = \frac{2}{3}$.

Probabilities can be written as either fractions or decimals. They always take values from 0 to 1. The probability of an event happening can be shown on the **probability scale**.

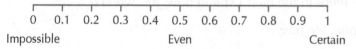

Probabilities of events not occurring

If one outcome is the absolute opposite of another outcome, such as 'raining' and 'not raining', then the probabilities of the two outcomes add up to 1.

Example 2

What is the probability of *not* scoring a number under 5 when you roll a dice?

From example 1, you know that P(number less than 5) is $\frac{4}{6}$.

Then, P(number not less than 5) $= 1 - \frac{4}{6}$

$$= \frac{6-4}{6} = \frac{2}{6} = \frac{1}{3}$$

Example 3

The probability that a woman washes her car on Sunday is 0.7. What is the probability that she does not wash her car on Sunday?

These two outcomes are the opposite of each other, so the probabilities add up to 1.

The probability that she does not wash her car is:

$1 - 0.7 = 0.3$

Example 4

A girl plays a game of tennis. The probability that she wins is $\frac{2}{3}$. What is the probability that she loses?

The probability of her not winning, or P(losing), is:

$1 - \frac{2}{3} = \frac{1}{3}$

Exercise 3A

1. A set of cards is numbered from 1 to 50.

   One card is picked at random. Give the probability that the number on it:

   a is even   b has a 7 in it   c has at least one 3 in it

   d is a prime number   e is a multiple of 6   f is a square number

   g is less than 10   h is a factor of 18   i is a factor of 50.

2. A bag contains 32 counters. Some are black and the others are white. The probability of picking a black counter is $\frac{1}{4}$.

   How many white counters are there in the bag? Explain how you worked it out.

3. Joe has 1000 tracks on his phone. He has:

   250 tracks of White rock

   200 tracks of Blues

   400 tracks of Country & western

   100 tracks of Heavy rock

   50 tracks of Quiet romantic.

He sets the player to play tracks at random.

What is the probability that the next track to play is:

a   White rock   **b**   Blues   **c**   Country & western

d   Heavy rock   **e**   Quiet romantic   **f**   not Heavy rock?

4   In each box of cereal there is a free gift of a model dinosaur.

There are five animals to make up the set.

You cannot tell which animal will be in a box. Each one is equally likely.

Tyrannosaurus Rex   Stegosaurus   Triceratops

Brachiosaurus   Diplodocus

a   Liam needs a diplodocus to complete his set.

    His sister Kiera needs a stegosaurus and a triceratops.

    They buy one box of cereal.

    i   What is the probability that the animal is a diplodocus?

    ii   What is the probability that the animal is a stegosaurus or a triceratops?

b   Their mother opens the box. She tells them the animal is not a brachiosaurus.

    i   Now what is the probability that the animal is a diplodocus?

    ii   Now what is the probability that the animal is a stegosaurus or a triceratops?

5   **a**   Aidan puts two white counters and one black counter in a bag.

    He is going to take one counter without looking.

    What is the probability that he will pick a black counter?

b   Aidan puts the counter back in the bag and then puts more black counters in the bag.

    Again, he is going to take one counter without looking.

    The probability that he will pick a black counter is now $\frac{2}{3}$.

    How many more black counters did Aidan put in the bag?

6 Fred has a bag of sweets that contains:

   3 yellow sweets

   5 green sweets

   7 red sweets

   4 purple sweets

   1 black sweet

He takes a sweet from the bag at random.

   a What is the probability that Fred will take a black sweet?

   b Copy and complete this sentence, writing in the correct colour of sweet.

The probability that Fred will not take a … sweet is $\frac{3}{4}$.

7 Look at this probability scale.

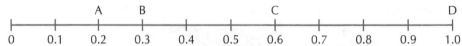

The probability of outcomes A, B, C and D are shown on the scale. Copy the scale and mark underneath it the probabilities of A, B, C and D *not* happening.

8 Copy and complete the table.

Outcome	Probability of outcome occurring (p)	Probability of event not occurring ($1 - p$)
A	$\frac{1}{4}$	
B	$\frac{1}{3}$	
C	$\frac{3}{4}$	
D	$\frac{1}{10}$	
E	$\frac{2}{15}$	
F	$\frac{7}{8}$	
G	$\frac{7}{9}$	

9 In a normal pack of playing cards there are 52 cards divided into four equal-sized sets: clubs (black), spades (black), diamonds (red) and hearts (red). The 13 cards within each set are numbered 1 (called the ace) to 10, plus a jack, a queen and a king.

A card is chosen at random from a pack of 52 playing cards. Calculate the probability that it is:

   a a black card   **b** an ace   **c** not an ace   **d** a diamond

   e not a diamond **f** not a 2   **g** not a picture   **h** not a king

   i not a red card **j** not an even number **k** not the ace of spades.

10 In a bus station there are 24 red buses, 6 blue buses and 10 green buses. Calculate the probability that the next bus to leave is:

   a green   **b** red   **c** red or blue   **d** yellow

   e not green   **f** not red   **g** neither red nor blue   **h** not yellow.

Activity: Will it happen? No it won't!

Design a spreadsheet to convert the probabilities of outcomes happening into the probabilities that they do not happen.

Use some of your answers to Exercise 3A, where you calculated some 'reverse probabilities', to help you to set up the spreadsheet and to test it to see if it works.

3.2 Mutually exclusive outcomes

Learning objective

• To recognise mutually exclusive outcomes

Key words
intersection
mutually exclusive
set
union
Venn diagram

Mutually exclusive outcomes are those that cannot occur together. Each excludes the possibility of the other happening. For example, when you roll a dice, throwing a 1 or a 6 are mutually exclusive as you cannot get both results on the same throw of one dice.

However, suppose you have a dice and are trying to throw numbers less than 4, but you also want to score an even number. Which number is common to both outcomes?

• The numbers less than 4 are 1, 2 and 3.
• The even numbers are 2, 4 and 6.

Because the number 2 is in both groups, the outcomes are not mutually exclusive. This means it is possible to achieve both outcomes at the same time if you throw a 2.

You can use a **Venn diagram** to illustrate this.

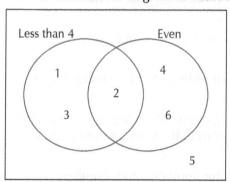

> **Hint** A set is just a collection of objects or numbers.

The two circles show the **sets** of the possible outcomes: 'less than 4' and 'even'.

In this diagram:

• the set 'less than 4' has 1, 2 and 3 in it
• the set 'even numbers' has 2, 4 and 6 in it

The numbers that satisfy either or both outcomes together are in the **union** of the sets.

The numbers that can satisfy either outcome are in the area where the sets overlap. This is the **intersection** of the sets. Because there are some numbers in the intersection, this shows that the outcomes are not mutually exclusive.

Notice that all of the possible outcomes of the dice roll are included in the Venn diagram. The number 5 does not form part of either outcome, so it is positioned outside the circles.

Once you have drawn a diagram like this, it shows you the probability of both outcomes happening together. This is when the outcomes in the intersection occur.

In this example you can see that only one number (2) belongs in both sets. That means the chance of rolling an even number that is also less than 4 is one of the six possible outcomes, so the probability is $\frac{1}{6}$.

Example 5

Liz is buying fruit. Here is a list of possible outcomes.

A: She chooses strawberries.

B: She chooses red fruit.

C: She chooses green apples.

D: She chooses red apples.

E: She chooses oranges.

She chooses one item only. State which pairs of outcomes are mutually exclusive.

a A and B   **b** A and E   **c** B and C   **d** B and D

  a Strawberries are red fruit, so they are not mutually exclusive.

  b Strawberries are not oranges, so they are mutually exclusive.

  c Green apples are not red fruit, so they are mutually exclusive.

  d Red apples are red fruit, so they are not mutually exclusive.

Exercise 3B

1 In a game you need to roll a dice and score an odd number larger than 2.

  a Draw a Venn diagram showing the two sets 'odd numbers' and 'numbers larger than 2'.

  b Are the outcomes 'scoring an odd number' and 'scoring a number larger than 2' mutually exclusive? Explain your answer.

  c Use your Venn diagram to state the probability of rolling an odd number larger than 2.

2 Soolin has a bag containing ten cards, each showing one of the integers 1 to 10. She is playing a game and needs to select a card at random that shows a prime number smaller than 4.

  a Draw a Venn diagram showing the two sets 'prime numbers' and 'numbers smaller than 4'.

  b Are the outcomes 'selecting a prime number' and 'selecting a number smaller than 4' mutually exclusive? Explain your answer.

  c Use your Venn diagram to state the probability of selecting a card showing a prime number smaller than 4.

3 A number square contains the numbers from 1 to 100.

1	2	3	4	5	6	7	8	9	10
11	12	13	14	15	16	17	18	19	20
21	22	23	24	25	26	27	28	29	30
31	32	33	34	35	36	37	38	39	40
41	42	43	44	45	46	47	48	49	50
51	52	53	54	55	56	57	58	59	60
61	62	63	64	65	66	67	68	69	70
71	72	73	74	75	76	77	78	79	80
81	82	83	84	85	86	87	88	89	90
91	92	93	94	95	96	97	98	99	100

Numbers are chosen from the number square. Here is a list of outcomes.

A: The number chosen is greater than 50.

B: The number chosen is less than 10.

C: The number chosen is a square number (1, 4, 9, 16, …).

D: The number chosen is a multiple of 5 (5, 10, 15, 20, …).

E: The number chosen has at least one 6 in it.

F: The number chosen is a factor of 100 (1, 2, 5, 10, …).

G: The number chosen is a prime number (2, 3, 5, 7, …).

State whether the outcomes in each pair are mutually exclusive or not.

a	A and B	**b**	A and C	**c**	B and C	**d**	C and D
e	B and F	**f**	C and F	**g**	C and G	**h**	D and E
i	D and G	**j**	E and F	**k**	E and G	**l**	F and G

4 A sampling bottle contains 40 different coloured beads.

a After 20 trials Dan has seen 12 black beads and 8 white beads. Does this mean that there are only black and white beads in the bottle? Explain your answer.

 Hint A sampling bottle is a plastic bottle in which only one bead can be seen at a time.

b You are told that there are 20 black beads, 15 white beads and 5 red beads in the bottle. State which of these pairs of outcomes are mutually exclusive.

i Seeing a black bead and seeing a white bead

ii Seeing a black bead and seeing a bead that is not white

iii Seeing a black bead and seeing a bead that is not black

iv Seeing any colour bead and seeing a red bead

5 a Tom has three coins in his pocket: a 1p, a 2p and a 5p.

If he takes out one or more coins, the possible outcomes are:

1 coin	a 1p or a 2p or a 5p
2 coins	a 1p and a 2p
	a 1p and a 5p
	a 2p and a 5p
3 coins	a 1p and a 2p and a 5p.

Write down the different totals of each of the seven different outcomes.

b Some money falls out of Tom's pocket. Which is more likely, that he loses more than 4p or less than 4p?

Explain your answer.

c Elle has 1p, 2p, 5p and 10p in her pocket.

As in part **a**, find the different totals that Elle could take out of her pocket.

d Some money accidently falls out of her pocket. Which is more likely, that she loses more than 10p or less than 10p?

Explain your answer.

6 The British coins currently are 1p, 2p, 5p, 10p, 20p, 50p, £1 and £2 coins.

Ilya has two different coins in his pocket.

a List all the possible different amounts of money that he could have in his pocket.

b Which is more likely, that he has more than 60p or that he has less than 60p?

Explain your answer.

7 a Copy and complete the table to show all the possible pairs of scores if you spin these two spinners.

Spinner 1	Spinner 2	Total score
+2	0	2
+2	−1	1

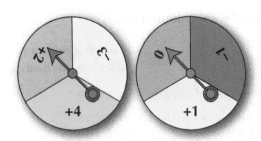

b Gary spins both spinners. Is he:

i more likely to get a positive total than a negative total

ii more likely to get an even total than an odd total?

Explain both your answers.

Challenge: Four men run a race

Imagine a race between two men called Arran and Benji. They could finish the race in two different ways:

- Arran first and Benji second (AB)
- Benji first and Arran second (BA).

A Now think about a race with Arran, Benji and Callum. How many ways can they finish the race?

B Extend this problem to four men, and so on. Put your results into a table. See if you can work out a pattern to predict how many different ways a race with ten men could finish.

C When you have finished this, explore what the factorial ⏺ button does on a calculator. This may help you to solve problems like this more quickly.

3.3 Using a sample space to calculate probabilities

Learning objective

- To use sample spaces to calculate probabilities

Key word

sample space

To help you work out the probabilities of events happening together you can use tables or diagrams called **sample spaces**.

A sample space is the set of all possible outcomes from a specific event.

Some events are simple, such as rolling a dice.

The sample space is {1, 2, 3, 4, 5, 6}.

Some are more complicated, such as rolling two dice.

The sample space could be written as {2, 3, 4, 5, 6, 7, 8, 9, 10, 11, 12}.

However, this is not helpful as the outcomes are not equally likely.

A better way to show this sample space is in a table that combines the equally likely outcomes of each dice.

		Dice 1					
		1	2	3	4	5	6
Dice 2	1	2	3	4	5	6	7
	2	3	4	5	6	7	8
	3	4	5	6	7	8	9
	4	5	6	7	8	9	10
	5	6	7	8	9	10	11
	6	7	8	9	10	11	12

This sample space shows the result of combining the outcomes from both dice. You can now see that the chances of rolling a total of 7 are far greater than the chance of rolling a total of 12.

Example 6

Find the probability of getting a head and a six when you roll a dice and toss a coin at the same time.

This sample space shows all the possible outcomes of throwing a coin and a dice together.

	1	2	3	4	5	6
Head	H, 1	H, 2	H, 3	H, 4	H, 5	H, 6
Tail	T, 1	T, 2	T, 3	T, 4	T, 5	T, 6

You can now work out the probability of getting both a head and a 6.

$$P(\text{outcome}) = \frac{\text{the number of ways the outcome could occur}}{\text{the total number of possible outcomes}}$$

P(head and a six) = $\frac{1}{12}$

You can also use a Venn diagram or a table to illustrate a sample space.

Example 7

In a class of 32 pupils, there are 15 boys. Out of eight left-handed pupils in the class, six are girls.

What is the probability of choosing a right-handed boy from the class?

You could put this information into a table.

	Boy	Girl	Total
Left-handed	2	6	8
Right-handed	13	11	24
Total	15	17	32

You can now see that:

P(right-handed boy) = $\frac{13}{32}$

You can also show this in a Venn diagram, like this one.

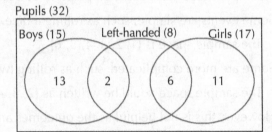

Pupils (32)

Exercise 3C

1. **a** Draw a sample space to show the results of rolling a coin and tossing a dice together.

    b Use your sample space to find the probability of scoring:

    i a 3 and a tail **ii** a head and an even number

    iii a number less than 5 and a tail.

2. A class has equal numbers of boys and girls. First one pupil is chosen, then another.

    a Write down the four possible combinations of pupils that could be chosen.

    b Jo says that the probability of choosing two boys is $\frac{1}{3}$. Explain why he is wrong.

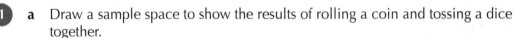

MR **3** A bag contains apples, bananas and pears. Liam chooses two fruits at random.

    a List all the possible outcomes.

    b You are told that you have more chance of choosing an apple and pear than an apple and a banana. Explain how that could happen.

4 A market trader sells jacket potatoes plain, with cheese or with beans. Clyde and Delroy each buy a jacket potato.

    a Copy and complete the sample space table.

Clyde	Delroy
plain	plain
plain	cheese

    b Give the probability of:

      i Clyde choosing plain

      ii Delroy choosing plain

      iii both choosing plain

      iv Clyde choosing plain and Delroy choosing beans

      v Clyde choosing beans and Delroy choosing cheese

      vi both choosing the same

      vii neither choosing plain

      viii each choosing a different flavour.

5 Bret rolls two dice and adds the scores together. Copy and complete the sample space of his scores.

	1	2	3	4	5	6
1	2	3				
2	3					

    a What is the most likely total?

    b Give the probability that the total is:

      i 4   **ii** 5   **iii** 1   **iv** 12   **v** less than 7

      vi less than or equal to 7   **vii** greater than or equal to 10   **viii** even

      ix 6 or 8   **x** greater than 5.

6 Bart rolls two dice and then multiplies the scores together.

    a Draw the sample space of his scores.

    b Which total is the most likely to occur?

    c What would be the probability of rolling a score greater than 17?

7 Bella spins two spinners together. Each spinner has the numbers 0, 1, 2 and 3 on it. What is the probability that the sum of the numbers she scores is less than 2?

8 A café makes 100 paninis. All the paninis are either meat or cheese, but 75 of them have salad in them, as well.

    30 paninis have cheese with salad.

    15 paninis have meat with no salad.

    a Show this information in a Venn diagram.

    b Use your Venn diagram to help you calculate the probability of selecting at random a cheese panini without salad.

Problem solving: Odd socks

In Andrew's sock drawer, all the socks are mixed up.

He knows that he has three pairs of blue socks, two pairs of white socks and a pair of green socks.

He takes two socks out without looking at their colour.

What is the probability that these socks are both the same colour?

Hint The answer is not $\frac{1}{3}$.

3.4 Experimental probability

Learning objective

• To calculate probabilities from experiments

How could you estimate the probability that a train will be late?

Will the train be late again today?

You could keep a record of the number of times that the train arrives late over a period of 10 days, and then use these results to estimate the probability that it will be late in future. The experiment here is observing the train, and the outcome you record is the train being late. The results enable you to find the **experimental probability** of the train being late.

$$\text{Experimental probability} = \frac{\text{number of trials that produce the outcome}}{\text{total number of trials carried out}}$$

Example 8

An electrician wants to estimate the probability that a new light bulb lasts for less than one month.

He fits 20 new bulbs and three of them fail within one month.

What is his estimate of the probability that a new light bulb fails within the month?

3 out of 20 bulbs fail within 1 month, so his experimental probability is $\frac{3}{20}$.

Example 9

A dentist keeps a record of the number of fillings she gives her patients over two weeks.

Here are her results.

Number of fillings	None	1	More than 1
Number of patients	80	54	16

Estimate the probability that a patient does not need a filling.

She saw 150 patients in total and 80 did not need fillings.

Experimental probability $= \frac{80}{150}$

$= \frac{8}{15}$

Example 10

A company manufactures items for computers. The number of faulty items is recorded in this table.

Number of items produced	Number of faulty items	Experimental probability
100	8	0.08
200	20	
500	45	
1000	82	

a Copy and complete the table.

b Which is the best estimate of the probability of an item being faulty? Explain your answer.

a

Number of items produced	Number of faulty items	Experimental probability
100	8	0.08
200	20	0.1
500	45	0.09
1000	82	0.082

b The best estimate is the last result (0.082), as the experiment is based on more results.

Exercise 3D

1 Jacob had an old computer that kept crashing. He kept a record of the days it crashed. This table shows his results.

Number of days	Number of days the computer crashed
50	32
100	72
150	106
200	139
250	175

  a From the results, would you say that on any one day there is a greater chance of Jacob's computer crashing or not? Explain your answer.

  b Which part of the table shows the most reliable data to use? Why?

  c From his data, how could Jacob estimate the probability of his computer crashing on any given day?

  d What would his answer be?

2 Amanda wants to test her octagonal dice to see if it is biased. She rolls the dice 100 times. Her results are shown in this table.

Score	1	2	3	4	5	6	7	8
Frequency	8	14	12	11	19	11	12	13

  a Do you think the dice is biased? Give a reason for your answer.

  b How could Amanda improve the experiment?

  c From the results, estimate the probability of her rolling a 7.

  d From the results, estimate the probability of her rolling a 3 or a 4.

  e From the results, estimate the chance of her *not* rolling a 5.

3 Faye started an experiment to find the probabilities of spinning a coin and getting a head or a tail. These are her results.

	Number of trials	Heads	Tails	P(H)	P(T)
First 20	20	8	12	$\frac{8}{20} = 0.4$	$\frac{12}{20} = 0.6$
Next 20	40	11	9	$\frac{19}{40} = 0.475$	$\frac{21}{40} = 0.525$

  a Use your own coin to spin the next 20, creating the next part of Faye's chart.

  b Complete the chart, writing down P(H) and P(T) from all 60 trials.

  c Repeat the above for the next 20 spins (giving a total of 80).

  d What do you notice about P(H)?

4 Lewis said to his dad: 'We always have chips and peas with our school lunch.'

His dad asked him to keep a record each day for a month of when chips and peas were on the menu.

He presented his results to his dad like this.

Week 1	Chips	No chips
Peas	2	1
No peas	1	1

Week 2	Chips	No chips
Peas	3	1
No peas	1	0

Week 3	Chips	No chips
Peas	2	0
No peas	1	2

Week 4	Chips	No chips
Peas	3	0
No peas	1	1

a Create a summary table to show Lewis's results.

b What is the probability of the school lunch including chips and peas?

c What is the probability of the school lunch including neither chips nor peas?

d What would make this a better method of sampling the school meals?

e Add another question:

5 Lightco have created a new halogen bulb. The inventors claimed that it would last for over 3000 hours. The customer research team wanted to test this theory out, so they ran a test, keeping some bulbs on continuously until they failed to light.

a How many weeks and days is 3000 hours?

b It was suggested that they test just 10 bulbs. Why is this not a good sample?

Finally, they decided to test 100 bulbs. These are the results.

Time, T (hours)	$0 \leqslant T < 500$	$500 \leqslant T < 1000$	$1000 \leqslant T < 1500$	$1500 \leqslant T < 2000$	$2000 \leqslant T < 2500$	$2500 \leqslant T < 3000$	3000+
Frequency	1	1	1	3	6	8	80

c What is the probability of one of the new halogen bulbs lasting over 3000 hours?

d Is the claim of the inventors true?

Problem solving: Roll the dice!

A Roll a pair of dice 50 times. Record the total of the two numbers shown, each time.

B What are the probabilities of each total being rolled?

C Repeat the experiment for another 50 rolls and record the results.

D Do you get the same results?

E Explain why your results are the same as – or different from – the first experiment.

Ready to progress?

I can calculate probabilities involving equally likely outcomes.
I can calculate probability from experimental data.

I know what mutually exclusive outcomes are.
I can use the probability of an outcome to calculate the probability that the outcome does not happen.
I can use sample spaces and Venn diagrams to help calculate probabilities.

Review questions

1 Helen chooses a numbered ball, at random, from the box. What is the probability that the number on the ball she chooses:

   a is a factor of 12   **b** is a square number

   c is the HCF of 24 and 36 **d** is a prime number?

(PS) **2** Matt draws all the triangles he can, so that:

   • one of the angles is always 70° smaller than another angle

   • all the angles are multiples of 10°.

   a   Sketch all the triangles Matt could have drawn, showing the angles.

   He drew each of his triangles in a different colour. One of the colours was red.

   b   What is the probability that his red triangle has:

      i   a right angle in it   **ii**   an angle of 30°?

   c   One of the other triangles was blue. What is the probability that this triangle was obtuse-angled?

(PS) **3** Kirsty drew all the rectangles she could with an area of 36 cm². The side lengths were all integer values.

   When she had finished she spilt coffee on just one of the rectangles. What is the probability that the rectangle she spilt coffee on had a perimeter longer than 36 cm?

4 Hannah had a set of cards.

   She used them to make fractions less than 1, like this.

a Create the sample space showing all the possible fractions Hannah could have made.

Hint She cannot use the same card twice in one fraction, for example, she cannot make:

b Hannah's little brother, Darren, picks up two of the cards at random and gives them to Hannah. What is the probability she can make a fraction equivalent to $\frac{1}{2}$ with them?

5 Lynne wanted to find out how many people used Facebook. She surveyed 100 people aged under 60 and 100 aged 60 or over.

Here are her survey results.

		Do you use Facebook?	
		Yes	No
Is your age	Under 60?	80	20
	60 or over?	10	90

a What percentage of people over 60 said they used Facebook?

b What is the probability of meeting someone under 60 who does not use Facebook?

6 Mr Speed teaches mathematics. He had a box that contained a cylinder, a cuboid, a cube, a square-based pyramid, a hexagonal prism, a cone and a sphere.

Katrina chose a shape at random out of the box.

Each shape was equally likely to have been chosen.

What is the probability that the shape Katrina chose:

a had six faces **b** had more than six faces?

7 Abbas was putting numbers on his spinner.

He wanted to arrange the numbers so that:

a he had more chance of spinning a negative number than not

b the probability of getting an even number was 0.5.

Draw a spinner showing numbers Abbas could have used.

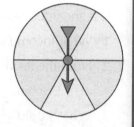

8 Brian regularly travelled down from Newcastle to London by train.

Over three weeks he counted how many times the drinks trolley passed him on each journey. His survey over three weeks showed these results.

		Monday	Tuesday	Wednesday	Thursday	Friday
Week 1	Newcastle to London	3	1	2	4	2
	London to Newcastle	2	2	3	1	0
Week 2	Newcastle to London	2	3	3	2	2
	London to Newcastle	3	4	4	1	1
Week 3	Newcastle to London	0	1	4	2	1
	London to Newcastle	2	2	3	4	2

a What is the modal number of times the drinks trolley passes?

b What is the experimental probability that on any journey the drinks trolley passes him four times?

Financial skills
Fun in the fairground

The fair has come to town.

Hoopla

You can buy five hoops for £1.25 and there are three types of prize.

You can win a prize by throwing a hoop over that prize and the base it is standing on!

Ben watched people at this stall. He counted how many tries they had and how many times someone won.

This table shows his results.

Prize	Number of throws	Number of wins
Watch	320	1
£10 note	240	4
£1 coin	80	2

Use the information about **Hoopla** to answer these questions.

1 What income would the throws that Ben recorded have made for the stall?

2 From the results shown, what is the probability of someone aiming for and winning:

   a a £1 coin   **b** a £10 note   **c** a watch?

3 What would you say is the chance of someone winning a prize with:

   a one hoop   **b** five hoops?

4 After watching, Ben decided to try for a £10 note. He bought 25 hoops and all his throws were aimed at the £10 note.

   a How much did this cost him?

   b What is the probability of his winning a £10 note?

5 On a Saturday afternoon, the stall would expect about 500 people each to buy a set of hoops. Assume that the throws would have been aimed at the various prizes in the same proportion as Ben observed.

   a How many of each prize would the stall expect to have to give away?

   b How much income would be generated from the 500 people?

   c If the watches cost £18 each, how much profit would the stall expect to make on a Saturday afternoon?

Hook a duck

On this stall, plastic ducks float in a moat around a central stall. Each duck has a number written on its underside, which cannot be seen until the duck is caught, by means of a hook on a stick. The number is checked by the stall holder.

If the number on the duck is:

1 you win a lollipop   2 you win a yo-yo   5 you win a cuddly toy.

Each time a duck is hooked, it is replaced in the water. Cindy, the stall holder, set up the stall one week with 45 plastic ducks.

🦆 Only one had the number 5 on it.

🦆 Nine had the number 2 on them.

🦆 All the rest had the number 1 on them.

Cindy charged 40p for one stick, to hook up just one duck.

Use the information about **Hook a duck** to answer these questions.

6 What is the probability of winning:

   a a cuddly toy   **b** a yo-yo   **c** a lollipop?

7 What is the probability of winning anything other than a lollipop?

8 Tom wanted his sister, Julie, to win a yo-yo.

   a How many ducks should Julie hook to expect to pick up at least one with a number 2 on it?

   b How much will it cost Tom to pay for the number of ducks he expects Julie to need, to win a yo-yo?

9 Before lunch on Sunday, Cindy took £100 from the stall.

   a How many ducks had been hooked that morning?

   b How many cuddly toys would you expect Cindy to have given away that morning?

   c How many yo-yos would you expect Cindy to have given away that morning?

10 Cindy bought the cuddly toys for £4 each and the yo-yos for 50p each. She gets the lollipops in a jar of 100 for £4. Cindy expects to take £250 on a Friday night.

   a How many ducks will she expect to be hooked that night?

   b How many lollipops will she expect to give away that evening?

   c How many yo-yos will she expect to give away that evening?

   d How many cuddly toys will she expect to give away that evening?

   e What will be the value of all the prizes she expects to give away that night?

4

Percentages

This chapter is going to show you:

- how to write one value as a percentage of another value
- how to use a multiplier to calculate a percentage increase or decrease
- how to write a change of value as a percentage increase or decrease
- how to use percentages to compare two quantities.

You should already know:

- the equivalence between fractions, decimals and percentages
- how to calculate a percentage of a quantity, with or without a calculator.

About this chapter

Increases and decreases are often given as percentages. You need to know how to interpret them. A pay rise of 4% does not mean that everyone receives the same amount of extra money. After a 4% pay rise, someone who started off earning £200 a week will earn an extra £8 a week. Someone who started off on £1000 a week will earn an extra £40 a week.

When you know the actual value of an increase or decrease it can be useful to calculate this as a percentage. For example, an increase of £2000 may be a lot in one context but not in another. If the price of a used car increases by £2000 from £5000 to £7000 you would call that a large increase. It is an increase of 40% of the original price. On the other hand, if the price of a house increases by £2000 from £200 000 to £202 000 you would call that a small increase. It is an increase of only 1% of the original price.

4.1 Calculating percentages

Learning objectives

- To write one quantity as a percentage of another
- To use percentages to compare quantities

Key word

percentage

You should be able to calculate a **percentage** of a quantity, using a calculator if necessary.

You may know two values and need to write one as a percentage of the other. This section will show you how to do that.

Example 1

Jon has a 500 g bag of flour. He uses 110 g in a recipe. What percentage of his flour has he used?

Write 110 as a fraction of 500 and change the fraction to a percentage.

$$\frac{110}{500} = \frac{22}{100} = 0.22 = 22\%$$

Because the numbers are straightforward, you should not need to use a calculator.

Example 2

Petra has £63.25 in her purse, but then spends £49.85. What percentage is that?

You can use the same method, even though the numbers are decimals.

The fraction is $\frac{49.85}{63.25} = 49.85 \div 63.25$

$\quad = 0.788\ldots = 78.8\ldots\%$

$\quad = 79\%$ to the nearest whole number

You could do the calculation all in one like this:

$49.85 \div 63.25 \times 100 = 79\%$

Exercise 4A

 1 These are some test marks. Write them as percentages.

a 20 out of 40	**b** 32 out of 40	**c** 9 out of 10
d 35 out of 50	**e** 17 out of 20	**f** 14 out of 25
g 19 out of 20	**h** 19 out of 25	**i** 19 out of 50

 2 Write these amounts of money as percentages.

a £3 out of £10	**b** £3 out of £20	**c** £3 out of £25
d £3 out of £30	**e** £3 out of £50	**f** £3 out of £100

3 Write these quantities as percentages.

    a 15 kg out of 20 kg   **b** 27 kg out of 50 kg   **c** 60 kg out of 80 kg

    d 400 g out of 1 kg   **e** 650 g out of 1 kg   **f** 900 g out of 2 kg

4 Write these quantities as percentages

    a 25 ml out of 250 ml   **b** 24 cm out of 40 cm   **c** 450 km out of 500 km

    d 21 days out of 30 days   **e** 200 years out of 1000 years   **f** 3 m out of 60 m

5 Melissa is running in a 10-kilometre race. She will complete 25 laps of a running track. What percentage of the race has she completed after she has run:

    a 1 lap   **b** 1 km   **c** 8 laps   **d** 8 km   **e** 2000 metres?

6 Write these quantities as percentages.

    a 4 mm out of 1 cm   **b** 40 cm out of 1 m   **c** 40 m out of 1 km

    d 600 ml out of 1 litre   **e** 290 g out of 1 kg   **f** 5 mm out of 5 cm

7 Write these amounts as percentages.

    a £33 out of £72   **b** £65 out of £264   **c** £68 out of £74

    d £6 out of £7   **e** £6 out of £70   **f** £123 out of £321

8 Write these amounts as percentages.

    a 25 kg out of 43 kg   **b** 2.5 kg out of 4.3 kg   **c** 9.4 kg out of 13.8 kg

    d 86 kg out of 249 kg   **e** 3000 kg out of 8200 kg   **f** 0.25 kg out of 0.68 kg

9 Jason earns £326. He pays £26 income tax.

    What percentage of his earnings does he pay in income tax?

10 These are Jordan's marks in three tests.

    English: 41 out of 60
    Maths: 59 out of 80
    Science: 84 out of 120

    a Write each mark as a percentage.

    b By comparing the percentages, decide which was his best mark.

11 Work out these amounts as percentages.

    a £3.49 out of £7.29   **b** £8.25 out of £9.45   **c** £8.25 out of £94.50

12 Alice starts the week with £524. She spends £182.50 on food. What percentage of her money is that?

13 A man travelling on an aircraft has a mass of 78.2 kg. His luggage weighs 19.3 kg. What percentage of the man's mass is the mass of his luggage?

14 Salt is a compound of sodium and chlorine.

    117 g of salt contains 46 g of sodium and the rest is chlorine.

    Find the percentage, by mass, of salt that is:

    a sodium   **b** chlorine.

15. A piece of chewing gum has a mass of 8 g.

    After it has been chewed for 5 minutes, its mass is 3.4 g.

    a What percentage of the original mass remains after 5 minutes?

    b What percentage of the original mass has been lost?

16. A magazine has 128 pages. Kerry estimates that there are 23 pages of adverts and 32 pages of photographs.

    Work out the percentages of:

    a adverts   **b** photographs

    in the magazine.

17. 6.40 litres of air contains 5.00 litres of nitrogen and 1.34 litres of oxygen.

    Work out the percentage of the air that is:

    a nitrogen   **b** oxygen   **c** other gases.

 18. There are 50 people in a room.

    a Ten more people come into the room. Gary says: 'The number in the room has increased by 20%.' Is Gary correct? Justify your answer.

    b Another ten people come into the room. Gary says: 'The number in the room has increased by another 20%.' Is Gary correct? Justify your answer.

Challenge: What is in the waste?

The Smith family and the Jones family weigh the waste they throw away. These are the results.

	Smith family	Jones family
Kitchen scraps (g)	1050	1150
Plastics (g)	270	175
Card or paper (g)	125	784
Other (g)	1820	1900

A Use percentages to compare the different types of waste thrown away by the two families.

B Draw a suitable diagram to illustrate your percentages.

4.2 Calculating percentage increases and decreases

Learning objective

• To use a multiplier to calculate a percentage change

In Book 1 you solved problems involving percentage increases and decreases. This section will show you an efficient way to do that by multiplying by a number, which is called a **multiplier**.

Example 3

The price of an article before tax is £64.50.

12% tax must be added.

Work out the price including tax.

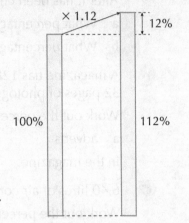

£64.50 is 100%. You need to add 12%.

100% + 12% = 112% altogether

112% = 1.12

£64.50 × 1.12 = £72.24

The price including the tax is £72.24.

You multiplied by 1.12, which is the multiplier in this example.

Example 4

The price of a pair of shoes is £64.50.

In a sale, the price is reduced by 12%.

Work out the sale price.

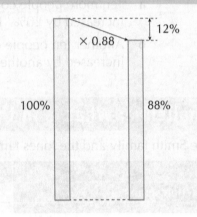

This time you need to subtract 12%.

100% − 12% = 88% = 0.88

The multiplier this time is 0.88.

£64.50 × 0.88 = £56.76

The sale price is £56.76.

Remember this formula:

> original value × multiplier = new value

Use the formula in this exercise.

Exercise 4B

(FS) **1** The price of a coffee machine is £86.00.

The price is going to increase by 20%.

    a What is the multiplier for a 20% increase?

    b Work out the price after the increase.

(FS) **2** Increase each of these prices by 20%.

    a £32.00   **b** £61.00   **c** £184.00   **d** £9.40

FS **3** **a** What is the multiplier for:
   i a 30% increase   **ii** a 36% increase
   iii a 43% increase   **iv** a 6% increase?

   b Increase £72 by:
   i 30%   **ii** 36%   **iii** 43%   **iv** 6%.

FS **4** Increase £42.00 by:
   a 15%   **b** 35%   **c** 85%   **d** 95%.

FS **5** 22% tax must be added to these prices.
   Work out the prices including tax.
   a £13.50   **b** £43.70   **c** £142.00   **d** £385.10

6 Increase 42 kg by:
   a 19%   **b** 39%   **c** 59%   **d** 92%.

7 Increase 270 cm by:
   a 1%   **b** 31%   **c** 51%   **d** 71%.

8 **a** What is the multiplier to decrease an amount by 15%?
FS **b** Decrease these amounts by 15%.
   i £63.00   **ii** £52.50   **iii** £262.00   **iv** £ 59.99

9 Work out the multiplier for a decrease of:
   a 10%   **b** 30%   **c** 37%   **d** 43%   **e** 75%.

FS **10** In a sale these prices are reduced by 40%.
   Find the sale price of:
   a a jacket reduced from £115
   b jeans reduced from £39
   c a shirt reduced from £29.50
   d boots reduced from £75.80.

   SALE
   40%
   OFF

FS **11** Work out the result of each reduction.
   a Reduce £37.60 by 35%   **b** Reduce 32 kg by 45%
   c Reduce 56 minutes by 32%   **d** Reduce 450 ml by 7%
   e Reduce 245 m by 3%   **f** Reduce 2500 hours by 95%

12 The price of a television is £360.

The price is increased by 10%. A month later it is reduced by 10%. Simon says:

The price must be £360 again.

Show that Simon is not correct.

13 The population of a town ten years ago was 57 000.

a It has decreased by 4% in the last ten years. Work out the population now.

b If it had increased by 4% in the last ten years, what would the population be now?

Financial skills: Percentage reduction

A A shop is thinking of reducing the prices of some items. There will be a percentage reduction.

Copy and complete this table to show the results of different percentage reductions.

Normal price (£)	Price after a 10% reduction	Price after a 30% reduction	Price after a 60% reduction
40			
75			
125			
350			

B A recent survey showed that some people think 30% off is better than 60% off.

Is this the case? Why might people think this?

4.3 Calculating a change as a percentage

Learning objective

• To work out a change in value as a percentage increase or decrease

In the last section, you used an efficient method to find the result of increasing or decreasing a value by a given percentage.

Sometimes you will know the increase or decrease in value and you will want to work out what percentage change this is, based on the original value.

Example 5

The number of students in a college increases from 785 to 834.

Work out the percentage increase.

You can use the formula: original value × multiplier = new value

785 × multiplier = 834   Divide by 785.

$\text{multiplier} = \frac{834}{785}$ The fraction is greater than 1 for an increase.

$= 1.062... = 106\%$ Use a calculator and round the answer.

This is the multiplier for an increase of 106% − 100% = 6%.

The value of 834 ÷ 785 was worked out on a calculator and rounded to two decimal places.

You can calculate percentage reductions in a similar way.

Example 6

In a sale, the price of a washing machine is reduced from £429 to £380.

Work out the percentage reduction.

This time:

429 × multiplier = 380   Divide by 429.

$\text{multiplier} = \frac{380}{429} = 0.885...$ The fraction is less than 1 for a decrease.

$0.885... = 89\%$ Round to the nearest whole number.

So this is a reduction of 100% − 89% = 11%.

Remember to use this formula:

> original value × multiplier = new value

Exercise 4C

1 The price of a computer increases from £360 to £378.

  a 360 × multiplier = 378

    Show that the multiplier is 1.05.

  b What is the percentage increase?

2 The mass of a baby increases from 3.5 kg to 5.6 kg.

  a 3.5 × multiplier = 5.6

    Work out the multiplier.

  b What is the percentage increase?

3 The number of people employed by a company increases from 163 to 230.

   a 163 × multiplier = 230

      Work out the multiplier. Round your answer to two decimal places.

   b What is the percentage increase?

4 The number of sheep on a farm is increased from 120 to 198.
   Work out the percentage increase.

5 The number of spaces in a car park is increased from 72 to 106.
   Work out the percentage increase.

FS **6** Tom gets a pay rise from £2452 per month to £2624 per month.
   Work out the percentage increase.

FS **7** The cost of building a new school increases from £12.3 million to £17.8 million.
   Work out the percentage increase.

FS **8** The price of a car decreases from £12 500 to £12 125.

   a 12 500 × multiplier = 12 125

      Work out the multiplier.

   b What is the percentage decrease?

9 The number of people in a health club decreases from 289 to 214.

   a 289 × multiplier = 214

      Work out the multiplier. Round your answer to two decimal places.

   b What is the percentage decrease?

FS **10** In a sale, prices are reduced by £40.

   Work out the percentage reduction for something marked down:

   a from £65 to £25   **b** from £119 to £79

   c from £242 to £202.

11 Companies are reducing their numbers of employees.
   Work out the percentage reduction when the number goes down:

   a from 38 to 28   **b** from 462 to 429   **c** from 1254 to 921.

FS **12** Mike is a builder. His annual income has decreased from £52 400 to £49 830.
   Work out the percentage decrease.

13 This table shows the numbers of people voting for each party in two local elections.

	2011 election	2014 election
Red Party	1281	1342
White Party	584	260
Blue Party	782	831

Work out the percentage change for each party. Say whether it is an increase or a decrease.

14 The total rainfall in a town in April is 20 cm.

    a In May there is 60% less rain than in April. What is the May rainfall?

    b In June there is 60% more rain than in May. What is the June rainfall?

15 Sam is earning £32 000 a year.

    a She gets a 4% pay rise. What is she earning now?

    b She gets another 4% pay rise. What is she earning now?

Problem solving: Five go on a diet

Five obese men join a slimming class.

Here are their masses.

Name	Jack	Oliver	Charlie	James	George
Mass (kg)	90	97	111	123	138

The class leader wants to set a target for them to aim for.

A If the target is to lose 5 kg, work out the target mass for each man.

B If the target is to lose 5%, work out the target mass for each man.

C Which do you think is a better target, 5 kg or 5%? Explain your answer.

Ready to progress?

I can write one value as a percentage of another value.
I can use percentages to compare two quantities.
I can use a multiplier to calculate a percentage increase or decrease on a value.
I can write a change of value as a percentage increase or decrease.

Review questions

1   Here are Isabelle's scores for three tests.

> Maths: 33 out of 40
> Science: 62 out of 75
> English: 71 out of 90

Change each mark to a percentage.

2   There are 463 men and 372 women in the audience at a concert.

Work out the percentage of the audience that are men.

(FS) 3   Olivia is buying a car for £3650.

She pays a deposit of £500.

What percentage of the price is the deposit?

(FS) 4   The cost of a train ticket is £53.20.

Prices are increased by 7%.

Work out the new price.

(FS) 5   The original price of a phone is £189.

SALE
Prices down
by 70%

Work out the sale price.

6   These are the heights of the highest mountains in Scotland, Wales and England.

Country	Scotland	Wales	England
Mountain	Ben Nevis	Snowdon	Scafell Pike
Height (m)	1344	1085	978

Write the heights of Snowdon and Scafell Pike as percentages of the height of Ben Nevis.

7   a   What percentage of 5^3 is 5?

  b   What percentage of 5^3 is 5^2?

8   Look at this sequence of numbers.

   60, 81, 102, 123, 144, 165, ...

  a   What is the term-to-term rule?
  b   What is the percentage increase from:
   i   60 to 81   ii   81 to 102

   iii   144 to 165   iv   165 to the next number?

9   The sides of a square are 20 cm long.

   The length of each side is increased by 1 cm.

  a   Work out the percentage increase in each side of the square.
  b   Work out the percentage increase in the perimeter of the square.
  c   Work out the percentage increase in the area of the square.

(FS)   10   Prices are being increased by $P\%$.

   This is a formula for increasing a price by $P\%$.

   New price $= \left(1 + \dfrac{P}{100}\right) \times$ old price

  a   Use this formula to increase a price of £250 by 7%.
  b   Show that this formula is the same as using a multiplier.

(FS)   11   Andy's income £1850 a month. His rent is £670 a month.

  a   What percentage of Andy's income does he spend on his rent?
  b   Andy's rent increases by 5%. What percentage of his income does he spend on rent now?

(FS)   12   A banker has a salary of £65 000.

  a   She is paid a bonus of £40 000. What percentage of her salary is that?
  b   She is given a pay rise of 16%. What is her new salary?

Challenge
Changes in population

1 This table shows the populations of the UK and Australia in different years. The numbers are in millions.

	1970	1980	1990	2000	2010
UK (millions)	55.7	56.3	57.3	58.9	62.3
Australia (millions)	12.5	14.7	17.1	19.2	22.1

a Find the percentage increase in the UK population from 1970 to 2010.

b Find the percentage increase in the Australian population from 1970 to 2010.

c Work out which decade had the greatest percentage increase in population in the UK.

d Work out which decade had the greatest percentage increase in population in the Australia.

e Show the figures in the table in a suitable diagram.

2 The population of the UK in 1901 was 38.2 million.

a Between 1901 and 1951 the population of the UK increased by 31.4%. Work out the population in 1951.

b If the population had increased by the same percentage between 1951 and 2001, what would the population have been in 2001?

c In fact, the population in 2001 was 59.1 million. What was the percentage increase between 1951 and 2001?

3 This table shows the changes in the world population. The numbers are in billions.

Year	1980	1990	2000
World population (billions)	4.44	5.26	6.07

Find the percentage increase:

a from 1980 to 1990　　b from 1990 to 2000　　c from 1980 to 2000.

4 In 1900 the world population was 1650 million and the population of Europe was 408 million.
In 1950 the world population was 2520 million and the population of Europe was 547 million.

a What percentage of the people in the world lived in Europe in 1900?

b What percentage of the people in the world lived in Europe in 1950?

5 This table shows the populations of different parts of the world in 2008 and estimated populations in 2025. The numbers are in millions.

	2008	2025
Asia (millions)	2183	2693
Africa (millions)	984	1365
Europe (millions)	603	659
South America (millions)	462	550
North America (millions)	444	514

a What percentage of the people in the world lived in Europe in 2008?

b What is the prediction of the percentage of the people in the world who will be living in Europe in 2025?

c Work out the predicted percentage increase in population between 2008 and 2025 for each region.

d Which region is predicted to have the largest actual increase in population? Is this the region with the largest predicted percentage increase?

6 Use your answers to questions 4 and 5 to comment on how the percentage of the people in the world who live in Europe has changed.

5

Sequences

This chapter is going to show you:

- how to use flow diagrams to generate sequences
- how to use the nth term for sequences
- how to work out the nth term of a sequence
- how to use the special sequence of Fibonacci numbers.

You should already know:

- how to create sequences and describe them, in words
- how to substitute numbers into an algebraic expression
- the sequence of square numbers
- the sequence of triangular numbers.

About this chapter

The Fibonacci numbers are nature's numbering system. They appear everywhere in nature, from the leaf arrangement in plants, to the pattern of the florets of a flower, the arrangement of the twists on a pinecone or the scales of a pineapple. On many plants, the number of petals is a Fibonacci number: lilies and irises have 3 petals, buttercups have 5 petals, delphiniums have 8 petals, corn marigolds have 13 petals and asters have 21 petals, whereas daisies can be found with 34, 55 or even 89 petals. The Fibonacci numbers are therefore applicable to the growth of every living thing, including a single cell, a grain of wheat, a hive of bees – and even how rabbits breed.

5.1 Using flow diagrams to generate sequences

Learning objective

• To use flow diagrams to generate sequences

Key words	
finite	flow diagram
generate	infinite

Flow diagrams are used in many branches of mathematics, science and computing to plan sequences of activities and to make decisions.

They can be used to give a list of instructions to **generate** a **finite** sequence.

• A finite sequence only has a fixed number of terms, such as the even numbers less that 21.

• An **infinite** sequence goes on for ever, such as the odd numbers.

This is an example of a flow diagram.

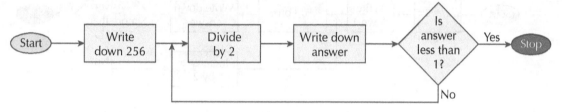

The rectangles are action boxes and the diamonds are decision boxes.

Example 1

Use the flow diagram above to generate a finite sequence.

The first term is 256.

The rule is 'Divide by 2'.

The sequence is 256, 128, 64, 32, 16, 8, 4, 2, 1, 0.5

Example 2

For each infinite sequence:

i describe how it is being generated

ii work out the next two terms.

a 2, 6, 10, 14, 18, 22, ...   **b** 1, 3, 27, 81, 243, ...

  a i The sequence is going up by 4.

    ii The next two terms are 26, 30.

  b i Each term is the previous term multiplied by 3.

    ii The next two terms are 729, 2187.

Exercise 5A 🖩

1. Use these flow diagrams to generate finite sequences.

a

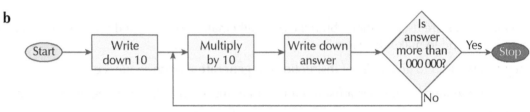

b

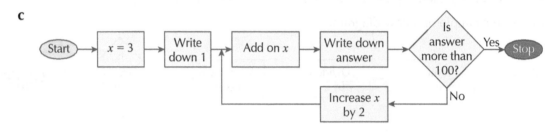

c

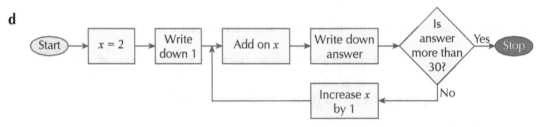

d

Start → $x = 2$ → Write down 1 → Add on x → Write down answer → Is answer more than 30? → Yes → Stop
No → Increase x by 1 (loops back to Add on x)

2. What do you call the numbers in the sequence generated by the flow diagram in Question **1b**?

3. What do you call the numbers in the sequence generated by the flow diagram in Question **1c**?

4. What do you call the numbers in the sequence generated by the flow diagram in Question **1d**?

5. Describe how each infinite sequence is generated.

   a 1, 4, 7, 10, 13, 16, ...　　　　**b** 1, 4, 16, 64, 256, 1024, ...

   c 1, 4, 8, 13, 19, 26, ...　　　　**d** 1, 4, 9, 16, 25, 36, ...

 6. Write down the first five terms for three different sequences beginning 1, 5, ..., ..., ... and explain how each of them is generated.

7. Describe how each of the following sequences is generated and write down the next two terms.

   a 40, 41, 43, 46, 50, 55, ..., ...　　　　**b** 90, 89, 87, 84, 80, 75, ..., ...

   c 1, 3, 7, 13, 21, 31, ..., ...　　　　**d** 2, 6, 12, 20, 30, 42, ..., ...

8 Draw a flow diagram to generate the first six terms for each set of instructions below.

    a Start at 1, multiply by 3   **b** Start at 2, multiply by 2

    c Start at 16, divide by 2   **d** Start at 4, subtract 3

PS **9** These patterns of dots generate sequences of numbers.

Write down the next four numbers in each sequence.

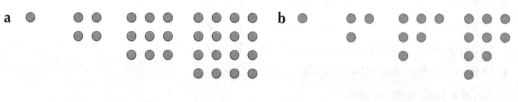

Problem solving: Algebraic flow diagrams

Draw flow diagrams, similar to those in Exercise **5A**, Questions **1c** and **1d**, to generate these finite sequences.

A 3, 8, 14, 21, 29, 38

B 40, 37, 32, 25, 16, 5

C 4, 6, 10, 18, 34, 66

5.2 The nth term of a sequence

Learning objective

• To use the nth term of a sequence

Key words	
algebraic expression	coefficient
constant term	nth term

You can describe a linear sequence by giving a rule for finding its terms. This rule is called the **nth term** rule and is an **algebraic expression**.

Look at this sequence.

5, 8, 11, 14, 17, 20, ...

The first term is 5 and you write this as $a = 5$.

The difference between the terms is 3 and you write this as $d = 3$.

The nth term for this sequence is $3n + 2$.

The number in front of n is called the **coefficient** of n and this is 3 in this example. The number by itself in the expression is called the **constant term** c and is 2 here.

Example 3

The nth term of the sequence 7, 11, 15, 19, 23, ..., is given by the expression $4n + 3$.

a Show this is true for the first three terms.

b Use the rule to work out the 50th term of the sequence.

   a When $n = 1$: $4 \times 1 + 3 = 4 + 3 = 7$ True ✓

      When $n = 2$: $4 \times 2 + 3 = 8 + 3 = 11$ True ✓

      When $n = 3$: $4 \times 3 + 3 = 12 + 3 = 15$ True ✓

   b When $n = 50$: $4 \times 50 + 3 = 200 + 3 = 203$

      So, the 50th term is 203.

Example 4

The nth term of a sequence is given by $1 - 3n$.

a Work out the first three terms of the sequence.

b Work out the 60th term of the sequence.

   a When $n = 1$: $1 - 3 \times 1 = 1 - 3 = -2$

      When $n = 2$: $1 - 3 \times 2 = 1 - 6 = -5$

      When $n = 3$: $1 - 3 \times 3 = 1 - 9 = -8$

      So, the first three terms are $-2, -5, -8$.

   b When $n = 60$: $1 - 3 \times 60 = 1 - 180 = -179$

      So, the 60th term is -179.

Example 5

The first six terms of a sequence are:

    9, 14, 19, 24, 29, 34, ...

What are the values of a and d in this sequence?

   The first term is 9, so $a = 9$.

   The difference between each term is 5, so $d = 5$.

Exercise 5B

1   Work out:

    i   the first three terms   **ii**   the 100th term

    for each sequence, when the nth term is given by:

    a $2n + 5$ **b** $4n - 1$ **c** $5n - 3$ **d** $3n + 2$ **e** $4n + 5$ **f** $10n + 1$

    g $\frac{1}{2}n + 2$ **h** $7n - 1$ **i** $\frac{1}{2}n - \frac{1}{4}$ **j** $10 - n$ **k** $20 - 2n$ **l** $7 - 3n$.

2 For each sequence, with the given nth term, write down:

  i the first four terms   **ii** the first term, a **iii** the difference, d.

  a $2n + 1$ **b** $2n + 2$ **c** $2n + 3$ **d** $2n + 4$

3 For each sequence, with the given nth term, write down:

  i the first four terms   **ii** the first term, a **iii** the difference, d.

  a $5n - 1$ **b** $5n + 2$ **c** $5n - 4$ **d** $5n + 3$

4 For each sequence, with the given nth term, write down:

  i the first four terms   **ii** the difference, d.

  a $3n - 1$ **b** $4n + 2$ **c** $6n - 4$ **d** $10n + 3$

 5 Look back at your answers to questions **2–4**.

What do you notice about the value of d and the coefficient of n?

6 For each sequence, write down the first term, a, and the difference, d.

  a 4, 9, 14, 19, 24, 29, …

  b 1, 3, 5, 7, 9, 11, …

  c 3, 9, 15, 21, 27, 33, …

  d 5, 3, 1, −1, −3, −5, …

7 Given the first term, a, and the difference, d, write down the first six terms of each sequence.

  a $a = 1, d = 7$ **b** $a = 3, d = 2$

  c $a = 5, d = 4$ **d** $a = 0.5, d = 1.5$

  e $a = 4, d = -3$ **f** $a = 2, d = -0.5$

 8 Look at these patterns, made from matchsticks.

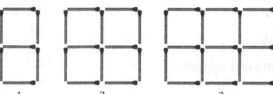

Pattern number	1	2	3
Number of matchsticks	7	12	17

  a How many matchsticks are there in the fourth pattern?   **This is the nth term.**

  b Write down the number of matchsticks in the nth pattern.

  c How many matchsticks are there in the 50th pattern?

Investigation: An *n*th term investigation

Here is a sequence.

5, 7, 9, 11, 13, …

For this sequence, $a = 5$ and $d = 2$. The *n*th term for the sequence is $2n + 3$.

In the expression $2n + 3$, the coefficient of *n* is 2 and the constant term, *c*, is 3.

Here are another five sequences and their *n*th terms.

a   8, 11, 14, 17, 20, …, $3n + 5$ b   8, 12, 16, 20, 24, …, $4n + 4$

c   7, 13, 19, 25, 31, …, $6n + 1$ d   4, 9, 14, 19, 24, …, $5n - 1$

e   2, 6, 10, 14, 18, …, $4n - 2$

A   Copy and complete the table below, using the sequences above. The first one has been done for you.

Sequence	First term, *a*	Difference, *d*	*n*th term	Coefficient of *n*	Constant term, *c*
5, 7, 9, 11, 13, …	5	2	$2n + 3$	2	3
a 8, 11, 14, 17, 20, …					
b 8, 12, 16, 20, 24, …					
c 7, 13, 19, 25, 31, …					
d 4, 9, 14, 19, 24, …					
e 2, 6, 10, 14, 18, …					

B   What is the connection between the difference, *d*, and the coefficient of *n*?

C   What is the connection between the difference, *d*, the first term, *a*, and the constant term, *c*?

5.3 Working out the *n*th term of a sequence

Learning objective

• To work out the *n*th term of a sequence

If you are given a linear sequence, how can you work out the *n*th term?

Here is a sequence you saw earlier.

5, 8, 11, 14, 17, 20, …, with $a = 5$ and $d = 3$

The *n*th term for the sequence is $3n + 2$. The coefficient of *n* is 3 and the constant term, *c*, is 2.

In the questions and in the investigation in the previous section, you should have discovered that the coefficient of *n* is the same as the difference, *d*. How can you work out the constant term, *c*?

In this sequence, the constant term is 2 and $a - d = 2$. So is it true that $c = a - d$?

Here is another sequence.

6, 11, 16, 21, 26, 31, …

The first term is 6 and the difference is 5, so $a = 6$ and $d = 5$.

The nth term for this sequence is $5n + 1$.

The coefficient of n (5) is the same as d and $c = a - d = 1$.

This rule works for all linear sequences.

Example 6

Work out the nth term for this sequence.

   7, 11, 15, 19, 23, 27, ...

   Here $a = 7$ and $d = 4$. So the coefficient of $n = 4$ and $c = 7 - 4 = 3$

   The nth term of the sequence is $4n + 3$.

Example 7

Work out the nth term for this sequence.

   40, 38, 36, 34, 32, 30, ...

   Here $a = 40$ and $d = -2$. So the coefficient of $n = -2$ and $c = 40 - (-2) = 40 + 2 = 42$.

   The nth term of the sequence is $-2n + 42$.

   You would normally write this as $42 - 2n$.

Example 8

Here is a pattern made from matchsticks.

a Work out the nth term. This is the number of matchsticks in the nth pattern.

b Work out the number of matchsticks in the 50th pattern.

   a The sequence for the number of matchsticks in the patterns is:

      3, 5, 7, 9, ...

      Here $a = 3$ and $d = 2$. So the coefficient of $n = 2$ and $c = a - d = 3 - 2 = 1$.

      The nth term is $2n + 1$.

   b In the 50th pattern, there are $2 \times 50 + 1 = 101$ matchsticks.

Exercise 5C

 1 Work out the nth term for each sequence.

     a 4, 10, 16, 22, 28, ...   **b** 9, 12, 15, 18, 21, ...

     c 9, 15, 21, 27, 33, ...   **d** 2, 5, 8, 11, 14, ...

     e 2, 9, 16, 23, 30, ...   **f** 8, 10, 12, 14, 16, ...

     g 10, 14, 18, 22, 26, ...   **h** 3, 11, 19, 27, 35, ...

     i 9, 19, 29, 39, 49, ...   **j** 4, 13, 22, 31, 40, ...

2 For each matchstick pattern below, work out:

i the number of matchsticks in the *n*th pattern

ii the number of matchsticks in the 50th pattern.

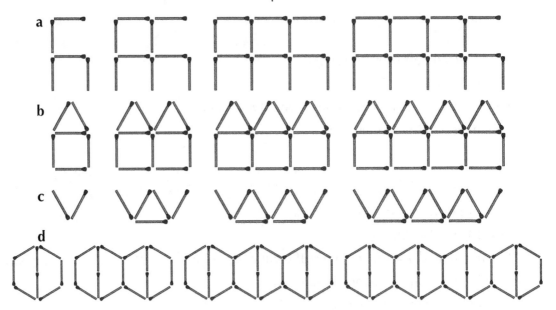

PS 3 Mia is using red and blue squares to make patterns.

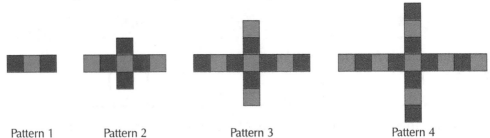

Pattern 1   Pattern 2   Pattern 3   Pattern 4

a Copy and complete the table.

Pattern	Number of blue squares	Number of red squares
1		
2		
3		
4		

b How many blue squares are there in her *n*th pattern?

c How many red squares are there in her *n*th pattern?

d How many squares are there altogether in her *n*th pattern?

4 Work out the *n*th term for each sequence.

a 90, 85, 80, 75, 70, …   b 43, 36, 29, 22, 15, …

c 28, 25, 22, 19, 16, …   d 44, 36, 28, 20, 12, …

5 Work out the *n*th term for each decimal sequence.

a 2.5, 3, 3.5, 4, 4.5, …   b 10.5, 13, 15.5, 18, 20.5, …

c 3.1, 3.2, 3.3, 3.4, 3.5, …   d 8, 7.8, 7.6, 7.4, 7.2, …

6 The pattern below is made from matchsticks of two different colours.

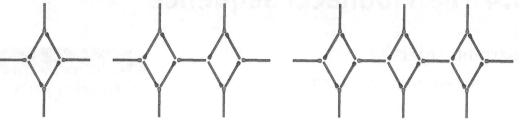

a Work out the *n*th term for:

  i the number of red-tipped matchsticks

  ii the number of blue-tipped matchsticks

  iii the total number of matchsticks.

b Use the *n*th term to work out, for the 20th pattern:

  i the number of red-tipped matchsticks

  ii the number of blue-tipped matchsticks

  iii the total number of matchsticks.

7 Sanna is saving money. She saves £4 in the first week, £7 in the second week, £10 in the third week and so on.

a How much does she save in the 10th week?

b How much does she save in the *n*th week?

8 This flow diagram can be used to generate sequences.

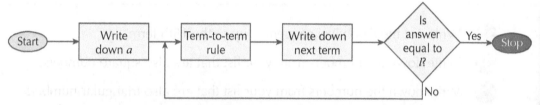

Write down a first term, *a*, a term-to-term rule and a value for *l* that you can use in the flow diagram to generate a sequence with five terms, so that:

a each term of the sequence is even **b** each term of the sequence is odd

c the sequence is the five times table **d** the sequence is the triangular numbers

e the numbers in the sequence all end in 1 **f** the sequence has alternating odd and even terms

g the sequence has alternating positive and negative terms.

Mathematical reasoning: Square sequences

Work out the *n*th term for each sequence.

A 1, 4, 9, 16, 25, ...

B 2, 5, 10, 17, 26, ...

C 0, 3, 8, 15, 24, ...

D $\frac{1}{2}$, 2, $4\frac{1}{2}$, 8, $12\frac{1}{2}$, ...

E 2, 6, 12, 20, 30, ...

5.4 The Fibonacci sequence

Learning objective

* To know and understand the Fibonacci sequence

Key word

Fibonacci sequence

An Italian mathematician, Leonardo Fibonacci, first wrote about this sequence in the 13th century.
The sequence is:

    1, 1, 2, 3, 5, 8, 13, ..., ...

Each term after the first two is the sum of the previous two terms.

Example 9

Work out the next four terms of the **Fibonacci sequence**.

$13 + 8 = 21$, $21 + 13 = 34$, $34 + 21 = 55$, $55 + 34 = 89$

So the next four terms are 21, 34, 55 and 89.

Exercise 5D

1 Write out the Fibonacci sequence, up to the 15th term.

2 Write down the numbers from your list that are also square numbers.

3 Write down the numbers from your list that are also triangular numbers.

4 Write down the numbers from your list that are prime numbers.

5 Work out the next four terms of each sequence.

    a 2, 2, 4, 6, 10, 16, ...   **b** 3, 3, 6, 9, 15, 24, ...

    c 1, 3, 4, 7, 11, 18, ...   **d** 2, 4, 6, 10, 16, 26, ...

 6 **a** **i** Sum the first three terms in the Fibonacci sequence.

    ii Sum the first four terms in the Fibonacci sequence.

    iii Sum the first five terms in the Fibonacci sequence.

    b Describe any patterns you see in your answers.

7 Write down any four consecutive terms from the Fibonacci sequence.

    i Multiply the first and the last together.

    ii Multiply the middle two together.

    iii What is the difference between the numbers?

    a Try this for at least two sets of four consecutive Fibonacci terms.

    b Does this rule work for any four consecutive Fibonacci terms?

8 **a** Divide each consecutive term in the Fibonacci sequence by the one before it. Copy and complete this table. The first two rows have been done for you.

Term	Previous term	Answer
1	1	1
2	1	2

Hint You may find a computer spreadsheet useful for this question.

b Explain what is happening.

Investigation: Steps and stairs

In most houses, there are 13 stairs in the staircases.

How many different ways are there of going up the stairs in a combination of one step or two steps at a time?

Take one stair. There is only one way of going up it (1).

Take two stairs. There are two ways of going up (1 + 1, 2).

Take five stairs. There are eight ways of going up them (1 + 1 + 1 + 1 + 1, 1 + 1 + 1 + 2, 1 + 2 + 1 + 1, 1 + 1 + 2 + 1, 1 + 2 + 2, 2 + 1 + 2, 2 + 2 + 1, 2 + 1 + 1 + 1).

A How many ways are there for three stairs?

B How many ways are there for four stairs?

C Copy and complete this table.

Number of stairs	Number of ways
1	1
2	2
3	
4	
5	8

D What do you notice about the numbers in the second column?

E Now work out how many ways there are to go up 13 stairs.

Ready to progress?

I can use flow diagrams to generate sequences.
I know how to work out the nth term of a sequence when given the rule.
I can write out the terms of the Fibonacci sequence.

I can work out the rule for the nth term of a sequence.

Review questions

1   This flow diagram can be used to generate sequences.

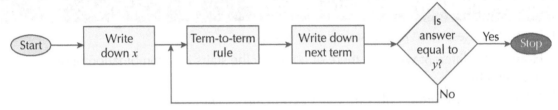

For example, if $x = 8$, the term-to-term rule is 'divide by 2' and $y = 0.25$, then the sequence is:

8, 4, 2, 1, 0.5, 0.25

Write down the sequences generated by:

	x	Term-to-term rule	y
a	3	Add 2	19
b	10	Subtract 5	−25
c	1	Multiply by −2	256
d	5000	Divide by 10	0.05

2   Given the first term a and the difference d, write down the first six terms of each sequence.

a $a = 3, d = 5$ b $a = 4, d = 2.5$ c $a = 10, d = -4$ d $a = -5, d = -3$

FS 3   The cost of house insurance increases by a fixed 5% of the initial cost each year.

This year the initial cost is £400.

How much will it cost:

a   next year   b   in 2 years' time   c   in 3 years' time

d   in 4 years' time   e   in n years' time?

4 Without working out the terms of the sequences, match each sequences to its nth term expression.

	Sequence	nth term
a	3, 9, 15, 21, 27, ...	$6n + 1$
b	10, 13, 16, 19, 22, ...	$3n - 1$
c	7, 13, 19, 25, 31, ...	$3n + 7$
d	2, 5, 8, 11, 14, ...	$6n - 3$

5 This is a pattern of huts made from matchsticks.

  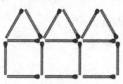

1 hut	2 huts	3 huts
6 matches	11 matches	16 matches

 Hint This is the nth term.

a Work out the number of matchsticks in the nth hut.

b Work out how many matchsticks will be needed for the 10th hut.

c Jack uses 101 matchsticks to make some huts.

How many huts does he make?

d Emily has 200 matchsticks.

What is the largest number of huts she can make?

How many matches will be left over?

 6 Here are pairs of sequences.

The second sequence is formed from the first sequence by a numerical rule.

a 5, 9, 13, 17, ... nth term is $4n + 1$ **b** 12, 18, 24, 30, ... nth term is $6n + 6$

   6, 10, 14, 18, ... nth term is ...   6, 9, 12, 15, ... nth term is ...

c 20, 17, 14, 11, ... nth term is $23 - 3n$ **d** 2, 7, 12 17, ... nth term is $5n - 3$

   16, 13, 10, 7, ... nth term is ...   4, 14, 24, 34, ... nth term is ...

i Write down the rule to get from the first sequence to the second sequence.

ii Work out the missing nth terms.

7 Work out:

i the first four terms   **ii** the 20th term

for each sequence, for which the nth term is given by:

a $n^2 + 1$ **b** $(n + 1)^2$ **c** $n^2 + 3$ **d** $n(n + 3)$ **e** $n^3 + 3$.

Investigation
Pond borders

Square stone paving slabs each have an area of 1 m² and are used to put borders around a square pond.

Here is an example for a square pond measuring 3 m by 3 m.

16 slabs fit around the pond.

How many slabs would fit around a pond measuring 5 m by 5 m?

What about a pond measuring 100 m by 100 m?

To solve this problem you need to follow these steps.

1 Break the problem into simple steps, using diagrams to help.

2 Set up a table to show the results.

3 Predict a rule and test it.

4 Use your rule to answer the questions.

Step 1 Draw simple diagrams for ponds of different sizes.

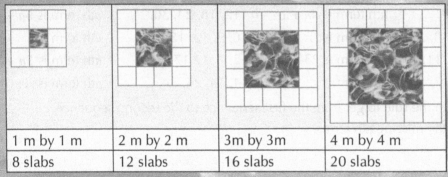

1 m by 1 m	2 m by 2 m	3m by 3m	4 m by 4 m
8 slabs	12 slabs	16 slabs	20 slabs

Step 2 Set up a table to show the results.

Pond size	Number of slabs
1 m by 1 m	8
2 m by 2 m	12
3 m by 3 m	16
4 m by 4 m	20

Step 3 Use the table to work out the *n*th term.

For the sequence 8, 12, 16, 20, ..., $a = 8$ and $d = 4$.

The *n*th term is $4n + 4$

So a 5 m by 5 m pond will need $4 \times 5 + 4 = 24$ slabs.

Draw a diagram to test your rule.

24 slabs are needed, so the rule works.

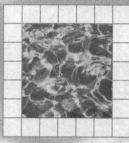

Step 4 Now you can use the rule to say that for a 100 m by 100 m pond,

$4 \times 100 + 4 = 404$ slabs will be needed.

Investigations

Make sure you follow the steps above and explain what you are doing clearly.

A Here are some rectangular ponds.

   2 m by 1 m   3 m by 1 m   4 m by 1 m   5 m by 1 m

  a Work out how many metre-square (1 m²) slabs fit around a 6 m by 1 m pond.

  b Work out how many metre-square slabs fit around an *n* m by 1 m pond.

  c Work out how many metre-square slabs fit around a 20 m by 1 m pond.

B Here are some different rectangular ponds.

   3 m by 2 m   4 m by 2 m   5 m by 2 m   6 m by 2 m

  a Work out how many metre-square slabs fit around a 7 m by 2 m pond.

  b Work out how many metre-square slabs fit around an *n* m by 2 m pond.

  c Work out how many metre-square slabs fit around a 20 m by 2 m pond.

6

Area of 2D and 3D shapes

This chapter is going to show you:

- how to work out the areas of triangles, parallelograms and trapezia
- how to work out the areas of compound shapes
- how to work out the surface areas of cuboids.

You should already know:

- how to work out the perimeters and areas of squares and rectangles
- how to work out the volumes of cubes and cuboids.

About this chapter

The Forth Railway Bridge crosses the Firth of Forth in the east of Scotland. The structure uses triangles to give the bridge strength.

It used nearly 58 000 tonnes of metal and 18 122 m^3 of granite and has a total length of 2528.7 metres.

We sometimes say a never-ending task is like 'painting the Forth Bridge'. This saying is based on the old – but false – belief that it takes so long to paint it that as soon as you finish one coat you have to start again.

A recent repainting involved applying 230 000 m^2 of paint, at a total cost of £130 million. This new coat of paint is expected to last at least 25 years, and possibly 40. Engineers had to work out the total surface area of the bridge to calculate how much paint was needed.

6.1 Area of a triangle

Learning objective

• To work out the area of a triangle

To work out the area of a triangle, you need to know the length of its **base** and its **height**. You measure the height by drawing a perpendicular line from the base to the angle above it. For this reason, it is sometimes called the **perpendicular height**.

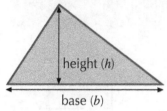

This diagram shows that the area of the triangle is half of the area of a rectangle that encloses the triangle.

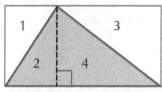

Area 1 = Area 2   and

Area 3 = Area 4

You can find the area of the rectangle by multiplying the base by the height.

So, the area of the triangle is:

$\frac{1}{2}$ × base × height

The formula for the area of a triangle is:

$A = \frac{1}{2} \times b \times h = \frac{1}{2}bh$

This is true for all triangles.

Example 1

Work out the area of this triangle.

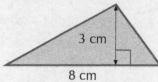

$A = \frac{1}{2} \times 8 \times 3 = 4 \times 3 = 12 \text{ cm}^2$

Example 2

Work out the area of this obtuse-angled triangle.

Notice that you have to measure the perpendicular height outside the triangle.

$A = \frac{1}{2} \times 6 \times 5 = 3 \times 5 = 15$ cm²

5 cm

6 cm

To work out the area of a **compound shape**, made from rectangles and triangles:

- work out the area of each part separately
- add together the areas of all the parts to obtain the total area of the shape.

Example 3

Work out the area of this compound shape.

4 cm

8 cm

10 cm

First divide the shape into a rectangle (A) and a triangle (B).

4 cm

8 cm   A   B

10 cm

Area of A = 8 × 4 = 32 cm²

Area of B = $\frac{1}{2} \times 6 \times 8 = 3 \times 8 = 24$ cm²

So the area of the shape = 32 + 24 = 56 cm².

Exercise 6A

 1 Work out the area of each triangle.

a
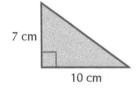
7 cm

10 cm

b
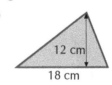
12 cm

18 cm

c
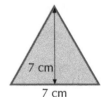
7 cm

7 cm

d
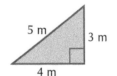
5 m   3 m

4 m

e
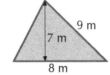
9 m

7 m

8 m

2 Work out the area of each triangle.

a
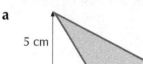
5 cm
6 cm

b
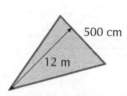
12 cm
10 cm

c
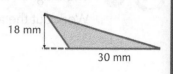
18 mm
30 mm

3 Work out the area of each triangle.

a
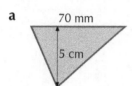
70 mm
5 cm

b
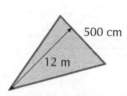
500 cm
12 m

c
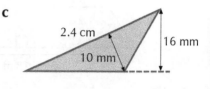
2.4 cm
16 mm
10 mm

4 Copy and complete the table for the triangles described in rows **a** to **e**.

Triangle	Base	Height	Area
a	6 cm	5 cm	
b	8 cm	7 cm	
c	11 m	5 m	
d	10 mm		40 mm²
e	12 m		42 m²

5 For each question, draw a coordinate grid on centimetre-squared paper, with axes for *x* and *y* both numbered from 0 to 6.

Then draw each triangle, with the given coordinates.

Work out the area of each triangle.

a Triangle ABC with A(2, 0), B(5, 0) and C(4, 4)

b Triangle DEF with D(1, 1), E(6, 1) and F(3, 5)

c Triangle PQR with P(2, 1), Q(2, 5) and R(5, 3)

d Triangle XYZ with X(0, 5), Y(6, 5) and Z(4, 1)

6 Work out the area of each compound shape.

a
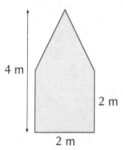
4 m
2 m
2 m

b
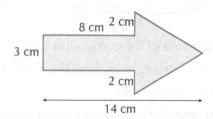
8 cm   2 cm
3 cm
2 cm
14 cm

c
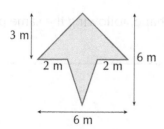
3 m
2 m   2 m   6 m
6 m

PS **7** The diagram shows the dimensions of a symmetrical flower garden.
Work out the area of the garden.

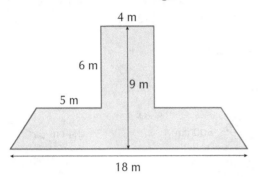

MR **8** This right-angled triangle has an area of 36 cm².

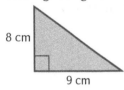

Find other right-angled triangles, with different measurements, that also have an area of 36 cm².

Investigation: Compound triangles

This compound shape is made from four different right-angled triangles.

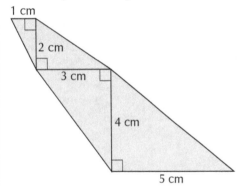

A Work out the area of the compound shape.

B The areas of the triangles making the shape form a sequence.

Write down this sequence.

What is the term-to-term rule for this sequence?

Write down the next number in this sequence.

C Another larger right-angled triangle is added to the shape, following the same pattern.

What is the area of the new compound shape?

6.2 Area of a parallelogram

Learning objective

* To work out the area of a parallelogram

Key word

parallelogram

To work out the area of a **parallelogram**, you need to know the length of its base and its perpendicular height.

These diagrams show that the parallelogram has the same area as a rectangle with the same base and height.

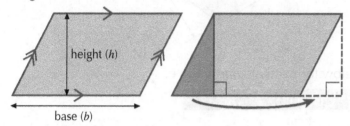
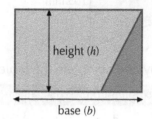

So the area of a parallelogram is:

base × height

The formula for the area of a parallelogram is:

$A = b \times h = bh$

Example 4

Work out the area of this parallelogram.

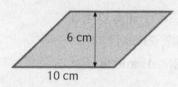

$A = 10 \times 6 = 60 \text{ cm}^2$

Exercise 6B

1 Work out the area of each parallelogram.

a
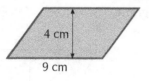

b
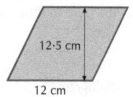

c
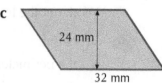

2 Work out the area of each parallelogram.

a
16 cm
5 cm
7 cm

b
7 m
7 m
8 m

c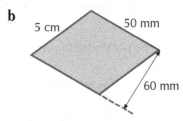
20 cm
4 cm
6 cm

3 Work out the area of each parallelogram.

a
8·4 cm
105 mm

b
5 cm
50 mm
60 mm

4 Copy and complete the table below for parallelograms **a** to **e**.

Parallelogram	Base	Height	Area
a	8 cm	4 cm	
b	17 cm	12 cm	
c	8 m	5 m	
d	15 mm		60 mm²
e	8 m		28 m²

5 For each question, draw a coordinate grid on centimetre-squared paper, with axes for x and y both numbered from 0 to 8.

Then draw each parallelogram with the given coordinates.

Work out the area of each parallelogram.

a Parallelogram ABCD with vertices at A(2, 0), B(6, 0), C(8, 5) and D(4, 5)

b Parallelogram EFGH with vertices at E(1, 2), F(4, 2), G(7, 7) and H(4, 7)

c Parallelogram PQRS with vertices at P(1, 8), Q(7, 5), R(7, 1) and S(1, 4)

6 The area of this parallelogram is 27 cm².

Work out the perpendicular height, h, of the parallelogram.

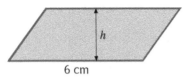
h
6 cm

7 The perpendicular height of a parallelogram is 8 cm and it has an area of 50 cm².

Work out the length of the base of the parallelogram.

8 Work out:

a the perimeter

b the area of this parallelogram.

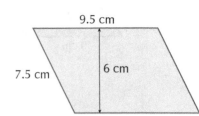
9.5 cm
7.5 cm
6 cm

9 Work out the value of *h* in this diagram.

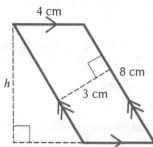

Challenge: Area of a rhombus

A Follow these instructions to work out a formula for the area of a rhombus.

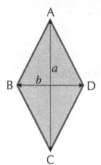

Hint The plural of rhombus is rhombi.

The lengths of the two diagonals of the rhombus are AC = *a* and BD = *b*.

Split the rhombus into two identical isosceles triangles.

By symmetry, the height of the triangle is $\frac{1}{2}a$.

The area of the triangle is $\frac{1}{2}b \times \frac{1}{2}a = \frac{1}{4}ab$.

The formula for the area of the rhombus is:

$A = 2 \times \frac{1}{4}ab = \frac{1}{2}ab$

B Use the formula to calculate the area of each rhombus.

a

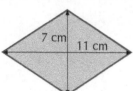

7 cm   11 cm

b

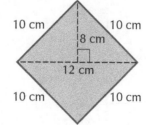

10 cm   10 cm
8 cm
12 cm
10 cm   10 cm

c

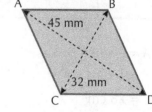

A   B
45 mm
32 mm
C   D

6.3 Area of a trapezium

Learning objective

- To work out the area of a trapezium

Key word

trapezium

To work out the area of a **trapezium**, you need to know the length of its two parallel sides, a and b, and the perpendicular height, h, between the parallel sides.

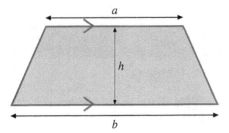

The diagram shows how you can fit two identical trapezia together to form a parallelogram.

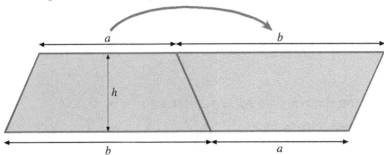

You know that the area of a parallelogram is given by base × height.

The area of the parallelogram formed from the two trapezia is $(a + b) \times h$.

But the area of one trapezium is half the area of the parallelogram.

The formula for the area of a trapezium is therefore given by:

$$A = \tfrac{1}{2} \times (a + b) \times h = \tfrac{1}{2}(a + b)h$$

Example 5

Work out the area of this trapezium.

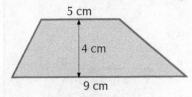

$A = \tfrac{1}{2} \times (9 + 5) \times 4$

$= \tfrac{1}{2} \times 14 \times 4$

$= 28 \text{ cm}^2$

Exercise 6C

1   Work out the area of each trapezium.

a 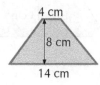   **b** 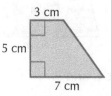   **c**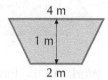

d 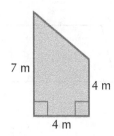   **e**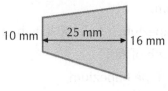

2   Copy and complete the table below for trapezia **a** to **f**.

Hint   Trapezia is the plural of trapezium.

Trapezium	Length, a	Length, b	Height, h	Area, A
a	4 cm	6 cm	3 cm	
b	10 cm	12 cm	6 cm	
c	9 m	3 m	5 m	
d	5 cm	5 cm		20 cm²
e	8 cm	12 cm		100 cm²
f	6 m		4 m	32 m²

3   The side of a swimming pool is a trapezium, as shown in the diagram.

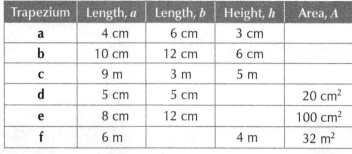

Work out its area.

4   Draw a coordinate grid on centimetre-squared paper, with axes for x and y both numbered from −6 to 6.

Plot the points A(3, 5), B(−4, 4), C(−4, −2) and D(3, −5).

Work out the area of the shape ABCD.

5   The diagram shows a shaded symmetrical trapezium drawn inside a square.

What is the area of the trapezium?

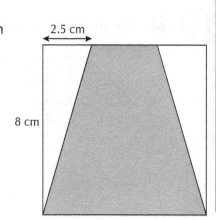

6 These three shapes have the same area.

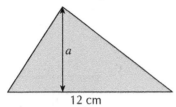 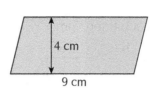 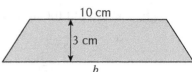

Work out the lengths marked *a* and *b*.

7 ABCD is a trapezium.

The height of the trapezium is half the length of AB.

CD is twice the length of AB.

Work out the area of the trapezium, when AB = 10 cm.

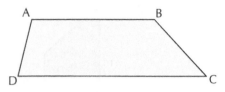

8 This is a pattern that is being used to make a patchwork quilt.

Each pattern is a square, measuring 12 cm by 12 cm square.

These are the measurements for each trapezium.

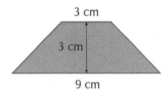

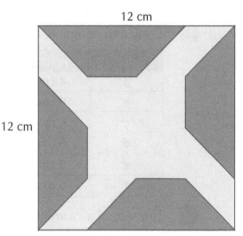

Work out the area of the yellow part of the pattern.

(MR) **9** The trapezium has an area of 80 cm². Work out the height, *h*.

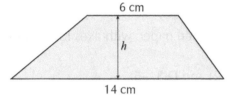

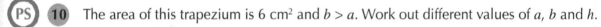

(PS) **10** The area of this trapezium is 6 cm² and *b* > *a*. Work out different values of *a*, *b* and *h*.

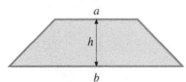

Problem solving: Pick's formula

The shapes below are drawn on a centimetre-squared grid of dots.

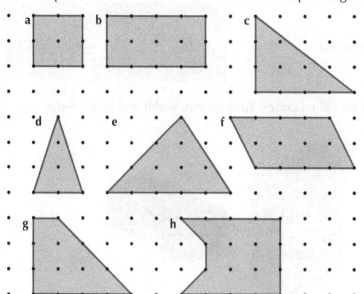

Shape	Number of dots on perimeter of shape	Number of dots inside shape	Area of shape (cm²)
a	8	1	4
b			
c			
d			
e			
f			
g			
h			

A Copy and complete the table, filling in the numbers for each shape. The first row has been done for you.

B Find a formula that connects the number of dots on the perimeter P, the number of dots inside, I, and the area, A, of each shape.

C Check your formula by drawing different shapes on a centimetre grid of dots.

6.4 Surface areas of cubes and cuboids

Learning objective

- To find the surface areas of cubes and cuboids

Shapes that are made from squares in 3D are called **cubes**. Their length, width and height (edge lengths) are all the same.

Shapes that are made from rectangles in 3D are called **cuboids**. Their length, width and height can all be different.

You can find the **surface area** of a cuboid by working out the total area of its six faces.

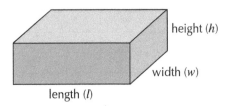

height (h)

width (w)

length (l)

Area of top and bottom faces = 2 × length × width = $2lw$

Area of front and back faces = 2 × length × height = $2lh$

Area of the two sides = 2 × width × height = $2wh$

So the surface area of a cuboid = $S = 2lw + 2lh + 2wh$.

Example 6

Work out the surface area of this cuboid.

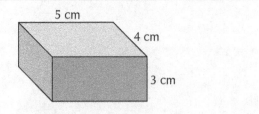

5 cm

4 cm

3 cm

The formula for the surface area of a cuboid is:

$S = 2lw + 2lh + 2wh$

$= (2 \times 5 \times 4) + (2 \times 5 \times 3) + (2 \times 4 \times 3)$

$= 40 + 30 + 24$

$= 94 \text{ cm}^2$

Example 7

Work out the surface area of this cube.

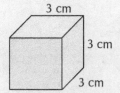

There are six square faces and each one has an area of 3 × 3 = 9 cm².

So the surface area of the cube is 6 × 9 = 54 cm².

Exercise 6D

1. Work out the surface area of each cuboid.

    a

    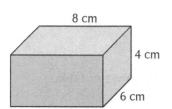

    b

    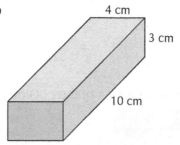

    c

    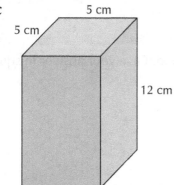

    d

    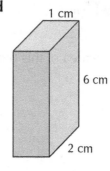

2. A cuboid measures 5 cm by 6 cm by 7 cm.

    Work out its surface area.

3 Work out the surface area of each cube.

a
4 cm
4 cm
4 cm

b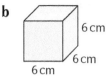
6 cm
6 cm
6 cm

c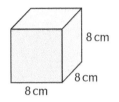
8 cm
8 cm
8 cm

d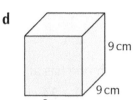
9 cm
9 cm
9 cm

4 Work out the surface area of a cube with an edge length of:

a 1 cm   b 5 cm   c 10 cm   d 12 cm.

5 Work out the surface area of the cereal packet.

23 cm
5 cm
18 cm

6 Work out the surface area of the outside of this open water tank.

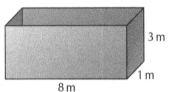
3 m
1 m
8 m

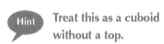
Hint Treat this as a cuboid without a top.

7 Work out the surface area of this block of wood. Give your answer in square centimetres.

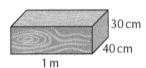
30 cm
40 cm
1 m

PS **8** Work out the surface area of this 3D shape.

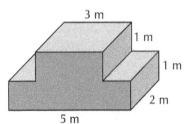
3 m
1 m
1 m
2 m
5 m

Investigation: An open box problem

An open box is made from a piece of card, measuring 18 cm by 14 cm, by cutting out a square from each corner.

Investigate the surface area of the open box formed when you cut out squares of different sizes from the corners of the card.

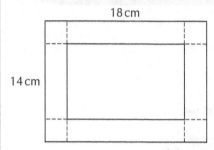

For example: cut off a square measuring 1 cm by 1 cm.

1 cm

1 cm

The area of the square is $1 \times 1 = 1$ cm^2.

So the area of the four squares is $4 \times 1 = 4$ cm^2.

The area of the card is $14 \times 18 = 252$ cm^2.

The surface area of the box is $252 - 4 = 248$ cm^2.

A   Copy and complete the table for squares of different sizes.

Size of squares cut out	Area of the four squares	Surface area of box
1 cm by 1 cm	4 cm^2	248 cm^2
2 cm by 2 cm		
3 cm by 3 cm		
4 cm by 4 cm		
5 cm by 5 cm		
6 cm by 6 cm		

B   Explain why you cannot cut out squares measuring 7 cm by 7 cm.

C   **a** Work out the differences between the surface areas as the size of the squares increases. Write these numbers as a sequence.

   b What is the term-to-term rule for this sequence?

   c Work out the nth term for this sequence.

Ready to progress?

I can use the appropriate formulae to find the area of triangles, parallelograms and trapezia.
I can use the formula $S = 2lw + 2lh + 2wh$ to work out the surface area of a cuboid.

Review questions

(FS) **1** **a** Weatbix cereals are sold in packs of two different sizes.

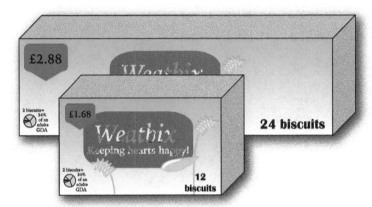

Which pack is better value for money?

b The smaller pack measures 20 cm by 10 cm by 5 cm.

Work out the total surface area of the larger pack, if its length is double that of the smaller pack, but all other measurements stay the same.

2 **a** Copy the centimetre-square grid below. Draw a right-angled triangle with an area of 12 cm².

Use the line AB for the base of the triangle.

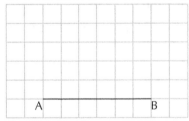

b Now draw an isosceles triangle with an area of 12 cm².

Use line AB for the base of the triangle.

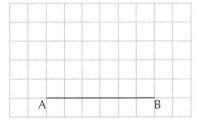

3   Work out the area of each shape.

a

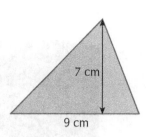

7 cm

9 cm

b

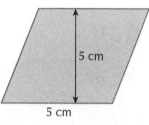

5 cm

5 cm

c

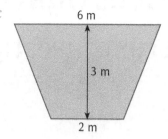

6 m

3 m

2 m

(PS)   4   This shape is made from four identical isosceles triangles.

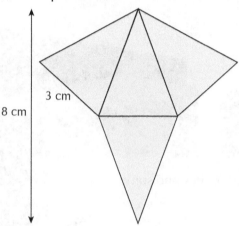

8 cm

3 cm

Work out the area of the shape.

(PS)   5   Work out the area of each symmetrical compound shape.

a

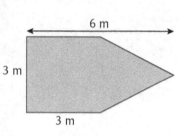

6 m

3 m

3 m

b

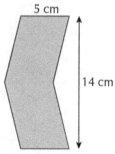

5 cm

14 cm

c

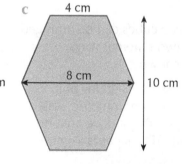

4 cm

8 cm

10 cm

(PS)   6   The volume of this box is 60 cm³.

The height is 1 cm more than the width and the length is
1 cm more than the height.

a   Work out the length, width and height of the box.
b   Work out the surface area of the box.

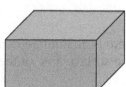

Investigation

A cube investigation

For this investigation you will need a collection of cubes and some centimetre isometric dotted paper.

Two cubes can only be arranged in one way to make a 3D shape, as shown.

Copy the diagram onto centimetre isometric dotted paper. The surface area of the shape is 10 cm².

Three cubes can be arranged in two different ways, as shown.

Copy the diagrams onto centimetre isometric dotted paper. The surface area of both 3D shapes is 14 cm².

Here is an arrangement of four cubes. The surface area of the 3D shape is 18 cm².

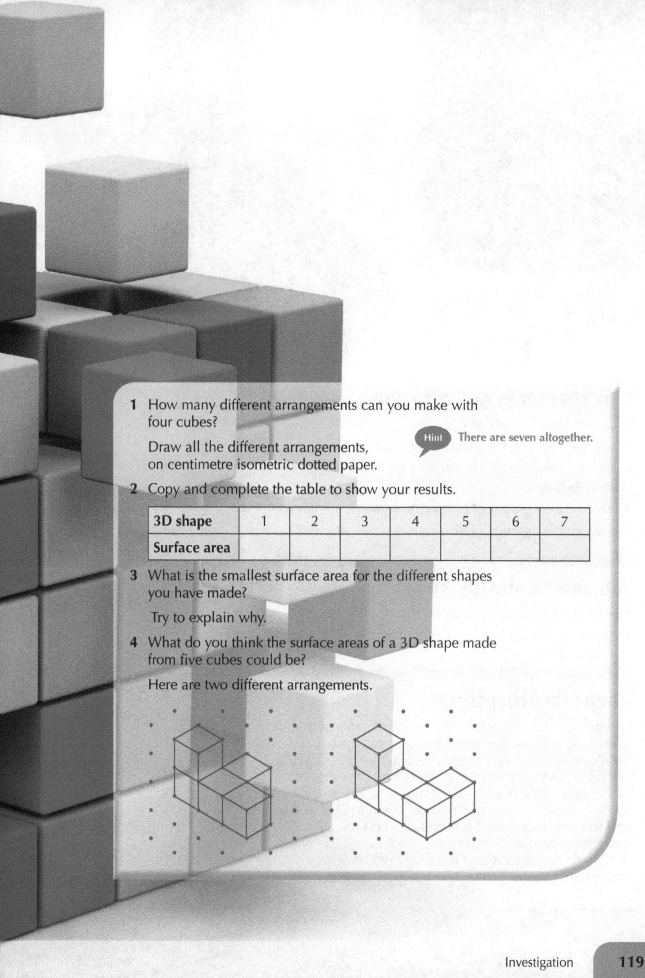

1 How many different arrangements can you make with four cubes?

Draw all the different arrangements, on centimetre isometric dotted paper.

Hint There are seven altogether.

2 Copy and complete the table to show your results.

3D shape	1	2	3	4	5	6	7
Surface area							

3 What is the smallest surface area for the different shapes you have made?

Try to explain why.

4 What do you think the surface areas of a 3D shape made from five cubes could be?

Here are two different arrangements.

7

Graphs

This chapter is going to show you:

- how to draw graphs of linear equations
- how to work out the gradient of a linear graph
- how to work out an equation of the form $y = mx + c$ from its graph
- how to draw graphs of simple quadratic equations
- how to draw graphs to illustrate real-life situations.

You should already know:

- how to plot coordinates in all four quadrants
- how to calculate with negative numbers.

About this chapter

Equations are a powerful mathematical tool used in design, engineering and computer software development. The fact that these equations can create graphs allows designers, engineers and developers to model real-life scenarios on computers. The shapes made by different equations allow them to be creative and to show movement on screen. To do this, the designers and engineers have to understand the nature of different equations and the effect of changing variables in them.

7.1 Graphs from linear equations

Learning objective

• To recognise and draw the graph of a linear equation

Key words

linear equation

variable

A **linear equation** connects two **variables** by a simple rule. Linear equations use any of the four operations: addition, subtraction, multiplication and division.

You have already met some examples of linear equations, such as:

• $y = x + 2$

• $y = 3x$

• $y = 2x + 1$

• $y = 8 - x$

and you have drawn graphs to represent them. The two variables x and y are connected by a simple rule each time.

Example 1

Draw a graph of the equation $y = 3x + 1$.

First, draw up a table of simple values for x. Then substitute that value of x in the equation to determine the corresponding y-value.

x	−2	−1	0	1	2
$3x$	−6	−3	0	3	6
$y = 3x + 1$	−5	−2	1	4	7

The middle line of the table helps you to work towards the final value of $3x + 1$.

Finally, take the pairs of (x, y) coordinates from the table, plot each point on a grid, and join up all the points.

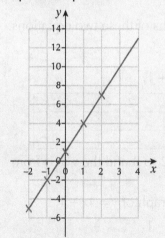

Notice that the line passes through other coordinates, as well. All of these fit the same equation, that is $y = 3x + 1$. Choose any points on the line that have not been plotted in the table and show that this is true.

1 **a** Copy and complete this table for the equation $y = x + 3$.

x	−2	−1	0	1	2	3
$y = x + 3$			3			

b Draw a coordinate grid, numbering the x-axis from −2 to 3 and the y-axis from −1 to 7.

c Use values from the table to draw, on your grid, the graph of $y = x + 3$.

2 **a** Copy and complete this table for the equation $y = x − 2$.

x	−2	−1	0	1	2	3
$y = x - 2$			−2			1

b Draw a coordinate grid, numbering the x-axis from −2 to 3 and the y-axis from −4 to 2.

c Use values from the table to draw, on your grid, the graph of $y = x − 2$.

3 **a** Copy and complete this table for the equation $y = 3x$.

x	−2	−1	0	1	2	3
$y = 3x$		−3	0			

b Draw a coordinate grid, numbering the x-axis from −2 to 3 and the y-axis from −6 to 12.

c Use values from the table to draw, on your grid, the graph of $y = 3x$.

d Copy and complete this table for the equations shown.

x	−2	−1	0	1	2	3
$y = x$	−2		0			3
$y = 2x$		−2	0		4	
$y = 4x$			0	4		

e Draw, on the grid you started with, the graph for each equation in the table.

(MR) **f** What two properties do you notice about each line?

g Use the properties you have noticed to draw the graphs of these two equations.

  i $y = 2.5x$ **ii** $y = 0.5x$

4 **a** Copy and complete this table for the equation $y = 4x + 1$.

x	−2	−1	0	1	2	3
$4x$		−4	0	4		
$y = 4x + 1$			1			

b Draw a coordinate grid, numbering the x-axis from −2 to 3 and the y-axis from −7 to 13.

c Use values from the table to draw, on your grid, the graph of $y = 4x + 1$.

5 **a** Copy and complete this table for the equation $y = 4x − 1$.

x	−2	−1	0	1	2	3
$4x$		−4	0	4		
$y = 4x - 1$			−1			

b Draw a coordinate grid, numbering the *x*-axis from −2 to 3 and the *y*-axis from −9 to 11.

c Use values from the table to draw, on your grid, the graph of $y = 4x - 1$.

6 a Copy and complete this table for the equations shown.

x	−2	−1	0	1	2	3
$y = x + 5$	3					8
$y = x + 3$		2			5	
$y = x + 1$			1	2		
$y = x - 1$			−1	0		
$y = x - 3$		−4			−1	

b Draw a coordinate grid, numbering the *x*-axis from −2 to 3 and the *y*-axis from −5 to 8.

c Draw the graph for each equation in the table above.

(MR) **d** What two properties do you notice about each line?

(PS) **e** Use the properties you have noticed to draw the graphs of these equations.
   i $y = x + 2.5$ **ii** $y = x - 1.5$

7 a Copy and complete this table for the equation $y = 2x + 4$.

x	−2	−1	0	1	2	3
$2x$	−4		0			6
$y = 2x + 4$			4			

b Draw a coordinate grid, numbering the *x*-axis from −2 to 3 and the *y*-axis from −8 to 10.

c Use values from the table to draw, on your grid, the graph of $y = 2x + 4$.

d Copy and complete the table below for the equations shown.

x	−2	−1	0	1	2	3
$y = 2x$		−2	0	2		
$y = 2x + 2$		0	2		6	
$y = 2x - 2$		−4	−2			
$y = 2x - 4$		−6	−4			

e Draw, on the grid you started with, the graph for each equation in the table.

(MR) **f** What two properties do you notice about each line?

(PS) **g** Use the properties you have noticed to draw the graphs of these two equations.
   i $y = 2x + 2.5$ **ii** $y = 2x - 2.5$

Challenge: Sloping graphs

A Draw, on the same set of axes, the graphs of these equations.
  • $y = 4x - 2$
  • $y = 4x + 2$

B Now draw, without any further calculations, the graphs of these equations.
  • $y = 4x - 0.5$
  • $y = 4x + 3.5$

7.2 Gradient (steepness) of a straight line

Learning objectives

- To work out the gradient of a graph from a linear equation
- To work out an equation of the form $y = mx + c$ from the graph

The **gradient** of a straight line is its steepness or slope. You can measure it by calculating the increase in value up the y-axis for an increase of 1 in the value along the x-axis.

Here are some examples of gradients.

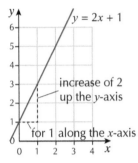

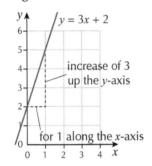

Hint The gradient of any straight line is calculated as the ratio of the change in the y-values to the change in the x-values, between two points on the line.

You will notice from these graphs that:

- the gradient is the same as the coefficient of x in the equation
- the line cuts the y-axis at the value that is added to the x-term.

The point where the line meets or cuts the y-axis is called the **y-intercept**.

A linear equation can be shown generally in the form:

$$y = mx + c$$

where m is the coefficient of x. The number added to the x-term is always the same so it is called the **constant** and is represented by c.

So, for any linear equation of the form $y = mx + c$:

- m is the same as the gradient of the line (steepness)
- c is the y-value where the line cuts through the y-axis.

Example 2

What is the equation of this graph?

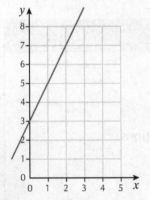

The gradient of the line is 2 and it cuts the y-axis at 3.

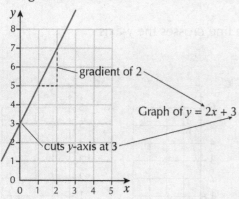

gradient of 2

Graph of $y = 2x + 3$

cuts y-axis at 3

Example 3

Write down the equation of the straight line that cuts the y-axis at (0, 5) and has a gradient of 3.

The equation of a straight-line graph can be written as $y = mx + c$ where m is the gradient and c is the y-intercept.

So the equation of this line is $y = 3x + 5$.

Exercise 7B

1 State the gradient of each line.

a 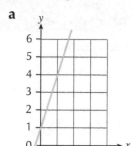   **b** 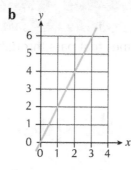   **c** 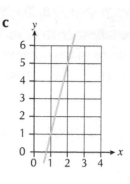   **d**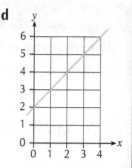

2 State the equation of the straight line with:

a a gradient of 3, passing through the y-axis at (0, 5)

b a gradient of 2, passing through the y-axis at (0, 7)

c a gradient of 1, passing through the y-axis at (0, 4)

d a gradient of 7, passing through the y-axis at (0, 15).

3 A straight line passes through the points (0, 1) and (4, 9).

a What are the coordinates of the point where this line cuts the y-axis?

b By plotting both points on a coordinate grid, calculate the gradient of the line.

c What is the equation of the line?

(4) For each graph:

    i   find the gradient of the coloured line

    ii write down the coordinates of where the line crosses the y-axis

    iii write down the equation of the line.

a 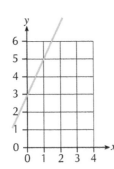 **b** 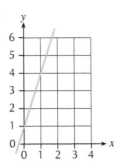 **c** 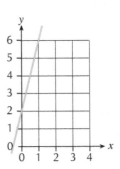 **d**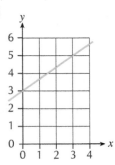

(5) A straight line passes through the points (0, 4) and (3, 10).

    a What are the coordinates of the point where this line cuts the y-axis?

    b By plotting both points on a coordinate grid, calculate the gradient of the line.

    c What is the equation of the line?

(6) A straight line passes through the points (1, 2) and (2, 5).

    a By plotting both points on a coordinate grid, calculate the gradient of the line.

    b What are the coordinates of the point where this line cuts the y-axis?

    c What is the equation of the line?

(PS) (7) What is the equation of the line that passes through the coordinates:

    a (0, 2) and (1, 5)   **b** (1, 3) and (2, 7)   **c** (2, 5) and (4, 9)?

Challenge: Lines through points

Write down the equations of ten different lines that pass through each point.

a (0, 4)   **b** (1, 1)   **c** (2, 5)

7.3 Graphs from simple quadratic equations

Learning objective

- To recognise and draw the graph from a simple quadratic equation

Key word

quadratic

In a **quadratic** equation one of the variables is squared.

Here are some examples of simple quadratic equations.

- $y = x^2$

- $y = x^2 + 1$

- $y = 3x^2$

- $y = 4x^2 - 3$

You can follow the same technique of finding coordinates that fit the equation and plotting them on a graph. This time, however, the lines are not straight!

Example 4

Draw the graph of the equation $y = x^2$.

First, draw up a table of values for x. Then substitute each value of x to determine the corresponding y-value.

x	−4	−3	−2	−1	0	1	2	3	4
$y = x^2$	16	9	4	1	0	1	4	9	16

Now take the pairs of (x, y) coordinates from the table and plot each point on a grid.

Join up all the points.

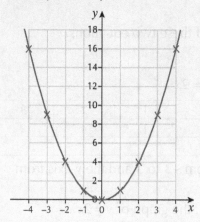

Notice the shape for a quadratic equation is a smooth curve. It is important to draw a quadratic graph very carefully, especially at the bottom of the graph where it needs to be a smooth curve.

Exercise 7C

1   a   Copy and complete this table for the equation $y = x^2 + 1$.

x	−3	−2	−1	0	1	2	3
x^2	9	4	1	0	1	4	9
$y = x^2 + 1$				1			

   b   Draw a coordinate grid, numbering the x-axis from −3 to 3 and the y-axis from −1 to 10.

   c   Use values from your table to draw, on your grid, the graph of $y = x^2 + 1$.

2   a   Copy and complete this table for the equation $y = x^2 + 3$.

x	−3	−2	−1	0	1	2	3
x^2							
$y = x^2 + 3$							

   b   Draw a coordinate grid, numbering the x-axis from −3 to 3 and the y-axis from −1 to 12.

   c   Use values from your table to draw, on your grid, the graph of $y = x^2 + 3$.

3 **a** Use your answers to questions 1 and 2 to help you copy and complete this table for the equations shown.

x	−3	−2	−1	0	1	2	3
$y = x^2 + 2$							
$y = x^2 + 4$							
$y = x^2 + 5$							

b Draw a coordinate grid, numbering the x-axis from −3 to 3 and the y-axis from −1 to 15

c Draw, on your grid, the graph for each equation in the table.

d What do you notice about each line?

e Use what you have noticed to draw the graphs of these two equations.

   i $y = x^2 + 2.5$ **ii** $y = x^2 − 1$

4 **a** Copy and complete this table for the equation $y = 2x^2$.

x	−3	−2	−1	0	1	2	3
x^2	9	4	1	0	1	4	9
$y = 2x^2$	18			0			

b Draw a coordinate grid, numbering the x-axis from −3 to 3 and the y-axis from −1 to 20.

c Use values from your table to draw, on your grid, the graph of $y = 2x^2$.

5 **a** Copy and complete this table for the equation $y = 3x^2$.

x	−3	−2	−1	0	1	2	3
x^2							
$y = 3x^2$	27			0			

b Draw a coordinate grid, numbering the x-axis from −3 to 3 and the y-axis from −1 to 30.

c Use values from your table to draw, on your grid, the graph of $y = 3x^2$.

6 **a** Use your answers to questions 4 and 5 to help you copy and complete this table for the equations shown.

x	−3	−2	−1	0	1	2	3
$y = 4x^2$							
$y = 5x^2$							
$y = 6x^2$							

b Draw a coordinate grid, numbering the x-axis from −3 to 3 and the y-axis from −5 to 60.

c Draw, on your grid, the graph for each equation in the table.

d What do you notice about each graph?

e Use what you have noticed to draw the graphs of these two equations.

   i $y = 0.5x^2$ **ii** $y = 2.5x^2$

7 Imagine that you are standing on a cliff, looking out to sea, on a clear day so that you can see the horizon. The distance (D km) to the horizon is related to the height (H metres) of your line of sight (your eyes) above sea level by the equation $H = D^2 + 13$.

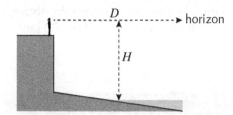

a   Draw a graph to show at what heights you can see distances out to sea from 0 to 50 km.

b   Find the distance you can see on a clear day from a height of 25 m above sea level.

c   On a clear day, Roy and Hayley were standing on the observation level of the Blackpool tower, which is at a height of 150 m above sea level. How far out to sea could they see?

7.4 Real-life graphs

Learning objective

• To draw graphs from real-life situations to illustrate the relationship between two variables

You find graphs everywhere, in newspapers, advertisements, on TV and the internet.

Most of these graphs show a relationship between two variables. One variable is shown on one axis and the other is shown on the other axis.

You can use a **distance–time graph**, like the one here, to describe a journey.

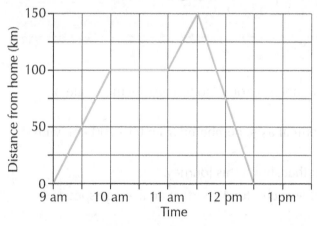

The vertical axis shows distance travelled and the horizontal one shows time spent travelling.

This graph shows that:

•   100 km was travelled in the first hour

•   then no distance at all was travelled for an hour (the line is flat)

•   the journey continued for half an hour, from 11:00 am to 11:30 am

•   the return journey started at 11:30 am.

Notice that the line shows the return journey by moving in the opposite direction. It goes down instead of up.

You can also use the graph to work out the average speed for a journey. In the example, above, you can see that the first 100 km were covered in 1 hour. This represents an average speed of 100 km/h. In general:

$$\text{average speed} = \frac{\text{total distance covered}}{\text{total time taken}}$$

Example 5

I set off from home to pick up my dog from the vet. I travelled $1\frac{1}{2}$ hours at an average speed of 60 km/h. It took me 30 minutes to get the dog settled into my car. I then travelled back home at an average speed of 40 km/h so that I didn't jolt the dog. Draw a distance–time graph of the journey.

Work out the key coordinates (time, distance from home).

- The start from home is represented by (0, 0).
- After $1\frac{1}{2}$ hours at 60 km/h I have travelled 90 km so I arrive at the vet's at $(1\frac{1}{2}, 90)$.
- I spend 30 minutes at the vet's without covering any distance, then I set off home at (2, 90).
- I travel 90 km home at 40 km/h so it takes $2\frac{1}{4}$ hours to get home. I arrive at $(4\frac{1}{4}, 0)$.

Now plot the points and draw the graph.

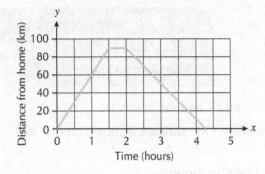

Exercise 7D

1. **a** Draw a grid with:
   - time on the horizontal axis, numbered from 0 to 5 hours, with a scale of 1 cm to 30 minutes
   - distance from home on the vertical axis, numbered from 0 to 100 km, with 1 cm to 20 km.

   b Draw on the grid the travel graph that shows this journey.
   - Marco travelled from home to Manchester Airport, at an average speed of 50 km/h. It took him 2 hours.
   - He stopped there for 30 minutes, and picked up his Auntie Freda.
   - He took her straight back home, driving this time at an average speed of 40 km/h.

    c Marco set off for the airport at 9:00 am. Use the graph to determine what time he arrived back home.

2. **a** Draw a grid with:
   - time on the horizontal axis, numbered from 0 to 2 hours, with a scale of 1 cm to 20 minutes
   - distance on the vertical axis, numbered from 0 to 60 km, with a scale of 1 cm to 10 km.

b Draw on the grid the travel graph that shows this journey.

- Sarah travelled to meet Dave, who was 60 km away.
- She left home at 11:00 am and travelled the first 40 km in 1 hour.
- She stopped for 30 minutes to buy a present, and then completed her journey in 20 minutes.

(MR) **c** Explain how to find Sarah's average speed over the last 20 minutes.

3 a Draw a grid with:

- time on the horizontal scale, numbered from 0 to 60 minutes, with a scale of 1 cm to 5 minutes
- depth on the vertical axis, numbered from 0 to 200 cm, with a scale of 2 cm to 50 cm.

b A swimming pool, 2 m deep, was filled with water from a hose. The pool was empty at the start and the depth of water in the pool increased at the rate of 4 cm/minute.

Copy and complete this table to show the depth of water after various times.

Time (minutes)	0	10	25	40	50
Depth (cm)			100		

c Draw a graph to show the increase in depth of water, against time.

(MR) **d** Explain how to use your graph to find the depth of water after $17\frac{1}{2}$ minutes.

4 a The swimming pool in question **3** was later filled with a different hose that poured in water more slowly, at the rate of 8 mm/minute. Draw a graph to show this.

(PS) **b** If the staff started to fill the pool at 1:00 pm, what time would it be full?

5 The water from another swimming pool was pumped out at the rate of 32 litres/minute. It took about 5 hours for the pool to be emptied.

a Copy and complete this table to show how much water is in the pool at various times.

Time (minutes)	0	50	100	150	200	250	300
Water left (litres)	9000						

b Draw a graph to show the amount of water left in the pool, against time.

(PS) **c** How long did it actually take to empty the pool?

Problem solving: Meeting in the middle

At 11 am, Billy and Leon set off towards each other from different places 32 km apart. Billy cycled at 20 km/hour and Leon walked at 5 km/hour.

A On the same grid, draw distance–time graphs of their journeys.

B Find out:

a the time at which they meet

b when they are 12 km apart.

Ready to progress?

I can complete a table of values for a linear equation and use this to draw a graph of the equation.
I can calculate the gradient of a straight line drawn on a coordinate grid.
I can work out an equation of the form $y = mx + c$ from its graph.
I can complete a table of values for a simple quadratic equation and use this to draw a graph of the equation.
I can draw and interpret graphs that illustrate real-life situations.

Review questions

1   The graph shows my journey in a lift. I got in the lift at floor number 8.

   a   The lift stopped at two different floors before I got to floor number 20. What floors did the lift stop at?

   b   For how long was I in the lift, while it was moving?

   c   After I got out of the lift at floor number 20, the lift went directly to the ground floor. It took 40 seconds.

      Copy the graph and show the journey of the lift from floor 20 to the ground floor.

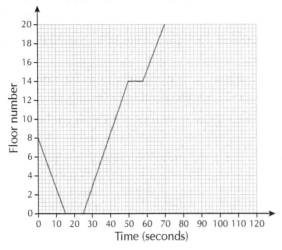

2   Look at the equation $y = 3x + 4$.

   a   When $x = 5$, what is the value of y?

   b   When $x = -5$, what is the value of y?

   c   Which of the equations below gives the same value of y for both $x = 5$ and $x = -5$?

      $y = 2x$ $y = 2 + x$ $y = x^2$

S **3** Rafael drew a graph to show the area of a circle for different radii.

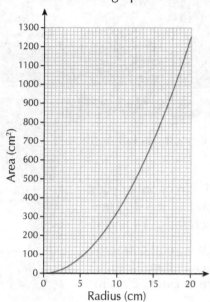

a Use the graph to find the radius of a circle with an area of 900 cm².

b What is the area of a circle with the radius of 12 cm?

4 The diagram shows a square drawn on a pair of axes.

The points A, B, C and D are at the vertices of the square.

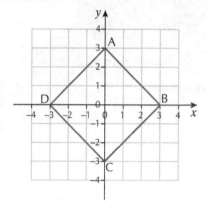

What is the equation of the line passing through:

a A and D   b   B and C

c A and B   d   D and C?

5 When a pendulum of length L metres is swinging, the approximate time, T seconds, of a complete swing is connected to L by the equation $L = \frac{1}{4}T^2$.

a Draw a graph showing this equation, for $T = 0$ to 4 seconds.

b Find the length of a pendulum that makes a complete swing in 1.5 seconds.

c Find the time taken to make a complete swing, for a pendulum of length 74 cm.

Challenge
The M25

The M25 motorway is an orbital motorway, 117 miles long, that encircles London.

It was built between 1973 and 1986.

For most of its length, the motorway has six lanes (three in each direction) or eight lanes (four in each direction). Between the busy junctions 12 and 14 it is 10 lanes wide, increasing to 12 lanes wide between the even busier junctions of 14 and 15.

It is one of Europe's busiest motorways. The road passes through several counties. Junctions 1–5 are in Kent, 6–14 are in Surrey, 15–16 are in Buckinghamshire, 17–24 are in Hertfordshire, 25 is in Greater London, 26–28 are in Essex, 29 is in Greater London and 30–31 are in Essex.

Use the information about the M25 to help you answer these questions.

1 How many years altogether did it take to build the M25?

2 a How many junctions are there on the M25?

b How many junctions are in Kent?

c Which county has the most junctions?

3 Use the scale shown.

a The point of the M25 that is nearest to the centre of London is Potters Bar. Approximately how far from the centre of London is this?

b The furthest point on the M25 from the centre of London is near junction 10. Approximately how far from the centre of London is this?

4 Tom used the M25 regularly and was so fed up with the slow traffic that he kept a diary of one day's journey from Junction 18 to 11. This is what he wrote down.

Junction 18 at 6:30 am. Steady 50 mph for 10 minutes. Crawling at 10 mph for the next 30 minutes. Stood still for 15 minutes in a jam. Made some progress at 30 mph for the next 15 minutes. Came off at junction 11.

Draw a distance–time graph to illustrate this journey.

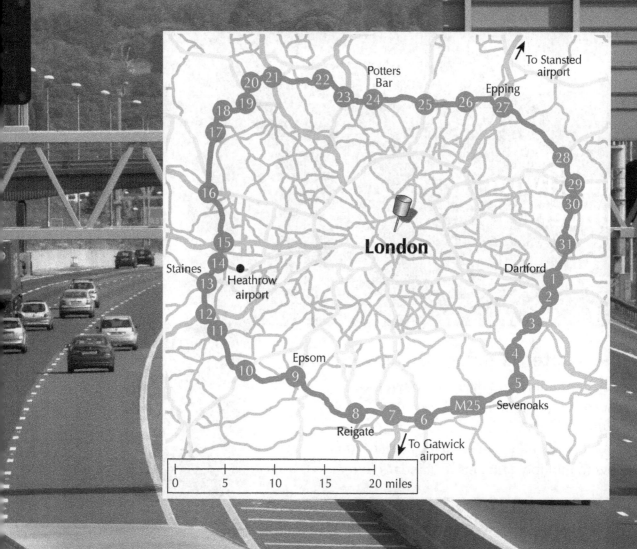

5　There are three airports near to the M25: Heathrow, Gatwick and Stansted. From Stansted to Heathrow via the M25 (anti-clockwise) is 67 miles. From Heathrow to Gatwick (anti-clockwise) is 43 miles. From Gatwick to Stansted (anti-clockwise) is 73 miles. A shuttle bus drives from Stansted, calls at Heathrow and Gatwick and then returns to Stansted. It does this four times a day.

　a　How far does it travel altogether?

　b　One day, it set off at 7:30 am from Stansted, arriving at Heathrow at 9:00 am. It stayed there for 20 minutes before setting of to Gatwick, arriving there 50 minutes later. After 20 minutes at Gatwick, it set off back to Stansted, taking 105 minutes. Draw a distance–time graph to illustrate this journey.

6　The legal speed limit on the M25 is 70 mph. Assuming that there are no hold-ups, how long would it take to drive around the M25 at the legal speed limit? Give your answer in hours and minutes.

> **Hint**　distance = speed × time

7　How long, to the nearest kilometre, is the M25? 5 miles ≈ 8 kilometres

8　The longest traffic jam on the M25 was 22 miles long. What percentage of the total length of the motorway was the length of the jam?

8

Simplifying numbers

This chapter is going to show you:

- how to multiply and divide by powers of 10
- how to round numbers to a specific number of significant figures
- how to write large numbers in standard form
- how to multiply numbers in standard form.

You should already know:

- how to multiply and divide by 10, 100 and 1000
- how to round to the nearest 10, 100 and 1000.

About this chapter

Our nearest star, Proxima Centauri, is 40 653 234 200 000 kilometres from Earth. An atom is 0.000 000 0001 metres wide.

When dealing with very large and very tiny numbers, it is easier to round them and present them as small numbers multiplied or divided by powers of 10. This is called standard form. It is a powerful tool that is widely used in science, especially by scientists such as astronomers who are dealing with massive spaces, or cell biologists, who are dealing with tiny ones. In this chapter, you will learn how to work with powers of 10.

8.1 Powers of 10

Learning objective

• To multiply and divide by powers of 10

Key word

power

In this section you will multiply and divide by **powers** of 10.

You have seen before how to write numbers with powers, for example:

• one hundred is $100 = 10^2$

• one thousand is $1000 = 10^3$

• ten thousand is $10\ 000 = 10^4$

• one hundred thousand is $100\ 000 = 10^5$

• one million is $1\ 000\ 000 = 10^6$.

Example 1

Work these out.

a 0.937×10 **b** 2.363×100 **c** $0.000\ 281 \times 10^4$ **d** $0.000\ 0742 \times 10^6$

a $0.937 \times 10 = 9.37$	Multiplying by 10 moves the digits 1 place to the left.
b $2.363 \times 100 = 236.3$	Multiplying by 100 (10^2) moves the digits 2 places to the left.
c $0.000\ 281 \times 10^4 = 2.81$	Multiplying by 10^4 moves the digits 4 places to the left.
d $0.000\ 0742 \times 10^6 = 74.2$	Multiplying by 10^6 moves the digits 6 places to the left.

Example 2

Work these out.

a $65 \div 100$ **b** $0.985 \div 10$ **c** $7.8 \div 10^3$ **d** $3.14 \div 10^6$

a $65 \div 100 = 0.65$	Dividing by 100 (10^2) moves the digits 2 places to the right.
b $0.985 \div 10 = 0.0985$	Dividing by 10 moves the digits 1 place to the right.
c $7.8 \div 10^3 = 0.0078$	Dividing by 10^3 moves the digits 3 places to the right.
d $3.14 \div 10^6 = 0.000\ 003\ 14$	Dividing by 10^6 moves the digits 6 places to the right.

In these examples, although it is correct to say that the digits move, it may look as if you are moving the decimal point.

Look at the examples again, carefully, and think about the number of zeros in the power of ten you are multiplying or dividing by. Compare that with the number of places you move the digits. What do you notice?

1 Multiply each number by 100.

   a 5.3   **b** 0.79   **c** 24   **d** 5.063   **e** 0.003

2 Divide each number by 100.

   a 83   **b** 4.1   **c** 457   **d** 6.04   **e** 34 781

3 Multiply each number by 10^3.

   a 6.43   **b** 0.685   **c** 35.2   **d** 8.074   **e** 0.0021

4 Divide each number by 10^3.

   a 941   **b** 5.23   **c** 568   **d** 0.715   **e** 45 892

5 Write down the answers.

   a 3.1×10 **b** 6.78×100 **c** 0.56×1000 **d** $34 \div 1000$

   e $823 \div 100$ **f** $9.06 \div 1000$ **g** 57.89×100 **h** $68.78 \div 100$

   i 0.038×1000 **j** $0.037 \div 10$ **k** $0.05 \times 10\ 000$ **l** $543 \div 100\ 000$

6 Write down the answers.

   a 4.25×10^3 **b** 5.67×10^2 **c** 0.451×10^3 **d** $23 \div 10^3$

   e $712 \div 10^2$ **f** $8.05 \div 10^3$ **g** 46.7×10^2 **h** $68.9 \div 10^2$

   i 0.027×10^3 **j** $0.049 \div 10^4$ **k** 0.06×10^4 **l** $432 \div 10^5$

7 The mass of one electron is 0.000 000 000 000 000 000 000 000 000 911 grams.

   a What is the mass of one million electrons?

   b What is the mass of one billion electrons?

MR **8** The mass of one atom of hydrogen is 0.000 000 000 000 000 000 000 001 673 8 g.

   a How many hydrogen atoms will there be in 16 738 g of hydrogen?

   b Approximately how many hydrogen atoms will there be in 1 kg of hydrogen?

PS **9** The closest Mars gets to Earth is approximately 30 000 000 miles.

   a How many hours would it take to travel there if the spacecraft travelled at:

      i 100 miles per hour   **ii** 1000 miles per hour?

   b How many days would it take to travel to Mars if the craft travelled at 3000 mph?

Investigation: Multiplying 9109

Look at this pattern.

$9109 \times 1 =$ $9109 \times 6 =$

$9109 \times 2 =$ $9109 \times 7 =$

$9109 \times 3 =$ $9109 \times 8 =$

$9109 \times 4 =$ $9109 \times 9 =$

$9109 \times 5 =$

A Copy the pattern and complete each line.

B Explain any patterns that you find.

8.2 Large numbers and rounding

Learning objective

* To round large numbers

Key words

approximate	estimate
round	

When you are discussing large quantities, you often only need to use an **approximate** number. To work this out, you **round** the number up or down to the nearest suitable figure.

You can also use rounded numbers to **estimate** the answers to questions.

Example 3

a The bar chart shows the annual income for a large company over five years. Estimate the income for each year.

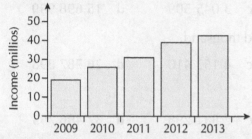

b The company chairman says: 'Income in 2013 was nearly 50 million pounds.' Is the chairman correct?

  a In 2009 the income was about 19 million pounds.

    In 2010 it was about 26 million pounds.

    In 2011 it was about 31 million pounds.

    In 2012 it was about 39 million pounds.

    In 2013 it was about 43 million pounds.

  b The chairman is wrong, as in 2013 the income is closer to 40 million pounds.

Example 4

The population of the United Kingdom is approximately 63 million. What are the smallest and largest population numbers this could represent?

  As the population is given to the nearest million, the actual population could be as much as half a million either way, so the population is between 63.5 million and 62.5 million.

  The smallest possible population would be 62 500 000.

  The largest possible population would be 63 499 999.

> **Hint** Notice that the population could not be 63 500 000, as you would round this up to 64 million.

Exercise 8B

1. The bar chart shows the populations of some countries in the European Community. Estimate the population of each country.

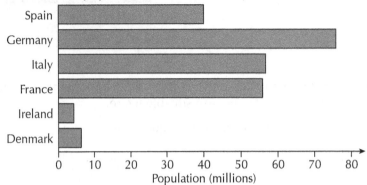

2. Round each number to the nearest ten thousand.

   a 3 547 812   b 9 722 106   c 3 045 509   d 15 698 999

3. Round each number to the nearest hundred thousand.

   a 4 678 923   b 8 631 095   c 4 153 410   d 26 787 898

4. Round each number to the nearest million.

   a 2 436 701   b 6 833 217   c 2 531 298   d 37 496 755

5. At one time, there were 2 452 800 people out of work. The government said: 'Unemployment is just over two million.' The opposition said: 'Unemployment is still nearly three million.' Who is closer to the real figure? Why?

6. There are estimated to be 6000 people living along the route of Hadrian's Wall. What are the highest and lowest actual numbers that this figure could represent?

7. There are estimated to be 8 million people living in London. What are the highest and lowest figures that the population of London could be?

(PS) 8. Three cruise ships docked in Dubrovnik. The residents were warned that one ship had about 3000 passengers, one ship had about 5000 passengers and one had about 1.5 thousand passengers. All the passengers from all three ships were expected in town that day.

   a What was the greatest number of passengers that the residents could expect to have in their town?

   b What was the lowest number of passengers that they could expect?

(MR) 9. In a show called 'Z-factor', two contestants, Mickey and Jenna, were told: 'Mickey received 8000 votes, Jenna received 7000 votes, but it was so close, there was only one vote between them.'

   Explain how this statement can be correct.

Investigation: Strange addition

Look carefully at this sum and how it is set out.

The units digit moves one space to the right as the sum works its way down—and the dots mean it goes on for ever!

A Write down the answer to this sum, as far as you think you can predict it.

B Try this again, just using even numbers as you work down, writing 2, 4, 6, 8, 10, ...

C Try this again, using multiples of 3 as you work down.

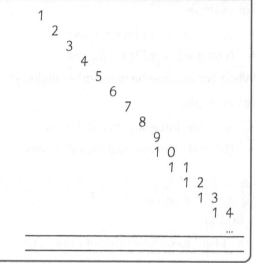

8.3 Significant figures

Learning objective

• To round to one or more significant figures

As well as rounding to the nearest 10, 100, 1000, you can also use **significant figures** to round numbers. For example, when you say: 'It's about 2 miles to school,' or: 'It took me 30 minutes to do my homework last night,' you are rounding to one significant figure (1 sf).

In any number:

• the digit with the highest place value is the most significant figure

• the digit with the next highest place value is the next most **significant**, and so on.

To round a number to one significant figure, you round the digit with the highest place value. If you round to two significant figures (2 sf) you use the digit after the one with the highest place value, and so on. For example, look at the number 382 649.

• The figure 3 has the highest place value, so it is the most significant. Rounded to one significant figure, 382 649 would be 400 000.

• The figure 8 is the next most significant. Rounded to two significant figures, 382 649 would be 380 000.

The method for rounding a number to significant figures is similar to the method for rounding to the nearest 10, 100, 1000 and so on.

• If the digit to the right of the significant number is 0, 1, 2, 3 or 4, then replace it by zero but leave the figure on the left unchanged.

• If this digit is 5, 6, 7, 8 or 9, then replace it by zero and add 1 to the figure before it, to the left.

A zero may or may not count as a significant figure, depending on its position.

When a zero is at the end of a whole number, it is not significant.

For example:

• 600 has one significant figure

• 720 has two significant figures.

When the number is less than 1, you start to count significant figures at the first non-zero digit. Any zeros that follow that digit count as significant.

For example:

- 0.007 has 1 significant figure
- 0.60 has 2 significant figures.

When zeros come between other digits, you count them as significant.

For example:

- 6.07 has three significant figures
- 0.070 01 has four significant figures.

Example 5

Round:

a 34.87 to one significant figure (1 sf)   **b** 10 942 to two significant figures (2 sf)

c 2.071 58 to three significant figures (3 sf)   **d** 2 315 489 to four significant figures (4 sf).

    a To round 34.87 to 1 sf

    The most significant digit is 3.

    The figure to the right of it is 4.

    The value of this is less than 5, so there is no rounding up. The 3 stays the same.

    Put in a zero to preserve the place value of the 3.

    Hence 34.87 ≈ 30 (1 sf).

    b To round 10 942 to 2 sf

    The two most significant digits are 10.

    The next figure to the right is 9.

    The value of this is greater than 4, so round up and add one to the 10, which makes 11.

    Replace the 9 and the other two figures with zeros to maintain the place values of the 1 and the 1.

    Hence 10 942 ≈ 11 000 (2 sf).

    c To round 2.071 58 to 3 sf

    The three most significant digits are 2.07.

    These are followed by 1.

    The value of this is less than 5, so the 2.07 stays the same.

    You do not need to add any zeros to preserve the place values of the digits in 2.07 as there are already three significant figures.

    Hence 2.071 58 ≈ 2.07 (3 sf).

    d To round 2 413 589 to 4 sf

    The four most significant figures are 2413.

    These are followed by 5, so round up and add one to the 2413, which makes 2414.

    Replace the 5 and the other digits with zeros to preserve the place value of the digits in the original number.

    Hence 2 413 589 ≈ 2 414 000 (4 sf).

Exercise 8C

1 State the number of significant figures in each number.

    a   1.325   **b**   320   **c**   5.24   **d**   0.509   **e**   8 million

2 Round each number to one significant figure.

    a   327   **b**   3760   **c**   60.8   **d**   0.9137

    e   0.0853   **f**   257   **g**   68.9   **h**   3650

    i   0.7396   **j**   9.52   **k**   583.2   **l**   0.084

3 Round each number to two significant figures.

    a   5329   **b**   49.7   **c**   9.053   **d**   752.2

    e   0.082 56   **f**   735.6   **g**   353   **h**   6492

    i   4880   **j**   94.4   **k**   0.869   **l**   0.005 57

4 Round each number to three significant figures.

    a   2148   **b**   9.5612   **c**   0.265 57   **d**   0.004 095 8

    e   0.000 716 94   **f**   25.857   **g**   268.26   **h**   10 945

    i   0.082 696   **j**   1.879 56   **k**   3 145 799   **l**   254 355

5 Round each number to the given number of significant figures.

    a   3.223 (2 sf)   **b**   7.5474 (1 sf)   **c**   13.56 (2 sf)   **d**   17.21 (3 sf)

    e   32.49 (1 sf)   **f**   36.99 (2 sf)   **g**   0.06371 (2 sf)   **h**   0.00754 (1 sf)

    i   83.697 (3 sf)   **j**   1.0097 (2 sf)   **k**   19.07 (1 sf)   **l**   23.14518 (4 sf)

6 Use a calculator to work out each division. Round the answer to the given number of significant figures.

    a   1 ÷ 3 (2 sf)   **b**   1 ÷ 7 (4 sf)   **c**   1 ÷ 13 (3 sf)   **d**   1 ÷ 11 (3 sf)

    e   11 ÷ 7 (2 sf)   **f**   12 ÷ 13 (3 sf)   **g**   17 ÷ 7 (4 sf)   **h**   21 ÷ 11 (3 sf)

    i   51 ÷13 (3 sf)   **j**   113 ÷ 37 (2 sf)   **k**   187 ÷ 32 (3 sf)   **l**   111 ÷ 33 (4 sf)

7 Tom painted his sitting room walls. He applied several coats of paint.

Altogether he used 7 litres of paint to cover an area of 75 m².

Work out the area, in square metres (m²), that was covered by each litre. Give your answer to two significant figures (2 sf).

8 The population of a small country is 2 647 100. Its total land area is 625 600 000 m².

Work out, correct to three significant figures, how many square metres (m²) there are to each person in this country.

9 Find the monthly pay, correct to four significant figures, of each person.

Name	Annual salary (£)
Mary	15 750
Chadrean	65 150
Sheila	55 590
Ismail	36 470
Sheena	45 800

10 Brant was planning to knock down a tall, circular chimney. He wanted to know approximately how many bricks there were in the chimney.

He counted about 215 bricks in one row, all the way round the chimney.

He counted, as best he could, 47 rows of bricks in the chimney, from bottom to top.

How many bricks would he expect to get after knocking the chimney down? Give your answer correct to one significant figure (1 sf).

Investigation: Patterns in calculations

A Use a calculator to work out the answer to each division.
Write down the full calculator answer.

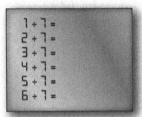

```
1 ÷ 7 =
2 ÷ 7 =
3 ÷ 7 =
4 ÷ 7 =
5 ÷ 7 =
6 ÷ 7 =
```

B Explain any patterns that you notice.

C Write down the answer to 3 ÷ 7 correct to 20 significant figures.

D In the answer to 3 ÷ 7, what will be the 100th significant figure?

8.4 Standard form with large numbers

Learning objective

• To write a large number in standard form

Key word

standard form

Standard form is a concise way of writing very large numbers, such as those that occur in IT and astronomy. For example, you can write 53 000 000 000 000 as 5.3×10^{13}.

Note that the power 13 is the number of places that the two digits (53) need to move to the right, to become 5.3.

There are two things to remember about a number expressed in standard form.

Rule 1: The first part is always a number that is greater than or equal to 1 but less than 10.

Rule 2: The second part is always a power of 10.

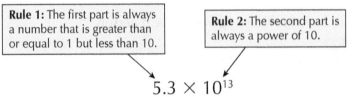

$$5.3 \times 10^{13}$$

Examples of numbers expressed in standard form include:

• the mean radius of the Earth, 6.4×10^6 m

• the speed of light, 2.998×10^8 m/s.

Scientific calculators display answers in standard form when the number has too many digits to fit on the display.

For example, the product of 6 000 000 and 7 000 000 is 42 000 000 000 000.

In standard form, this is 4.2×10^{13} so it could be displayed as:

   or

on the calculator.

Check to see how your calculator displays this answer.

Example 6

Express each number in standard form.

a 760 000   **b** 5420   **c** 36×10^7 **d** 0.24×10^5

  a Count how many places to move the digits to the right, so that you have a number between 1 and 10. For 760 000 that will be five places. This means that the power of 10 will be 5.

      $760\ 000 = 7.6 \times 10^5$

  b Count how many places to move the digits to the right, so that you have a number between 1 and 10. For 5420 that will be three places. This means that the power of 10 will be 3.

      $5420 = 5.42 \times 10^3$

  c Although 36×10^7 is written as a number multiplied by a power of 10, it is not in standard form, as the first part is not a number between 1 and 10.

    Work out the number in standard form, like this.

      $36 \times 10^7 = 3.6 \times 10 \times 10^7 = 3.6 \times 10^8$

  d 0.24×10^5 is not in standard form, as the first part is not a number between 1 and 10.

    Work out the number in standard form, like this.

      $0.24 \times 10 = 2.4$

      $10^5 = 10 \times 10 \times 10 \times 10 \times 10$

      So $0.24 \times 10^5 = 0.24 \times 10 \times 10 \times 10 \times 10 \times 10 = 2.4 \times 10 \times 10 \times 10 \times 10 = 2.4 \times 10^4$

      $0.24 \times 10^5 = 2.4 \times 10^4$

Example 7

Write each standard-form number as an ordinary number.

a 3.45×10^6 **b** 8.9×10^3 **c** 7.632×10^4

  a Move the digits to the left through the same number of places as the power of 10.

      $3.45 \times 10^6 = 3.45 \times 1\ 000\ 000 = 3\ 450\ 000$

  b Move the digits to the left through the same number of places as the power of 10.

      $8.9 \times 10^3 = 8.9 \times 1000 = 8900$

  c Proceed as in parts **a** and **b**.

      $7.632 \times 10^4 = 7.632 \times 10\ 000 = 76\ 320$

1 Write each number in standard form.

   a 5690   **b** 1 200 000   **c** 938 000   **d** 77 800

   e 396 500 000   **f** 561   **g** 73   **h** 4 300 000 000

2 Write each number in standard form.

   a 3.4 million   **b** 5.6 thousand   **c** 26 million   **d** 45 thousand

   e 258 million   **f** 547 thousand   **g** two hundred million   **h** half a million

3 In all English speaking nations, a billion is one thousand million. Write each number in standard form.

   a 8 billion   **b** 12 billion   **c** 150 billion   **d** a hundred billion

   e 6.7 billion   **f** 15.5 billion   **g** a thousand billion   **h** a billion billion

4 Write each standard-form number as an ordinary number.

   a 2.3×10^6 **b** 4.56×10^2 **c** 6.75×10^5 **d** 3.59×10^3

   e 9×10^6 **f** 2.01×10^6 **g** 3.478×10^4 **h** 8.73×10^7

   i 6.7×10^5 **j** 3.85×10^{10} **k** 7.8×10^8 **l** 5.39×10^9

5 Find the square of each number, giving your answer in standard form.

   a 500   **b** 4200   **c** 370   **d** 9000

   e 650   **f** 30 000   **g** 5 million   **h** 1.5 million

6 Write each number in standard form.

   a 73×10^6 **b** 256×10^2 **c** 77×10^5 **d** 259×10^3

   e 900×10^6 **f** 70.1×10^6 **g** 3478×10^4 **h** 1873×10^7

   i 0.7×10^5 **j** 0.85×10^{10} **k** 0.08×10^8 **l** 0.086×10^9

(PS) 7 When someone has a haircut, there will, on average, be approximately one million pieces of human hair left on the floor to be swept up. At Snippers, a popular hair stylist shop, about 300 people come in every week for a haircut. Approximately how many pieces of hair will Snippers sweep up each week? Give your answer in standard form.

(PS) 8 Together, 1.09×10^{27} electrons will have a total mass of 1 gram. How many electrons will have a total mass of 1 kg?

Activity: Astronomical numbers

Find the distance from the Sun to each planet in our solar system.

Write each distance both as a normal number and in standard form.

8.5 Multiplying with numbers in standard form

Learning objective

• To multiply with numbers in standard form

Now that you know how to write numbers in standard form, you can practise using them.

First, you need to know more about multiplying different powers of 10.

Example 8

Evaluate these numbers. **a** $10^2 \times 10^3$ **b** $10^3 \times 10^5$ **c** $10^6 \times 10^1$

a Write each power out in full.

$$10^2 \times 10^3 = (10 \times 10) \times (10 \times 10 \times 10)$$
$$= 10 \times 10 \times 10 \times 10 \times 10 = 10^5$$

> **Hint** You can simply add the powers of the tens when you are multiplying them.

b $10^3 \times 10^5 = 10^{(3+5)} = 10^8$ Apply the same rule as in part **a**.

c $10^6 \times 10^1 = 10^7$ Apply the same rule as in part **a**.

Example 9

Complete each calculation. Give your answer in standard form. Do not use a calculator.

a $(3 \times 10^3) \times (2 \times 10^4)$ **b** $(4 \times 10^2) \times (5 \times 10^3)$

a Rewrite the problem as $3 \times 2 \times 10^3 \times 10^4$.

Then multiply the numbers and add the powers of 10.

$$3 \times 2 \times 10^3 \times 10^4 = 6 \times 10^7$$

b As in part **a**, rewrite the problem without the brackets.

But multiplying 4 by 5 gives 20, which is not in standard form, so it needs to be converted.

$$4 \times 5 \times 10^2 \times 10^3 = 20 \times 10^5$$
$$20 \times 10^5 = 2 \times 10 \times 10^5 = 2 \times 10^6$$

Example 10

Light from the Sun takes about 8 minutes to reach the Earth. Light travels at 299 792 kilometres per second. How far is the Sun from the Earth? Give your answer in standard form, correct to three significant figures (3 sf).

First, convert 8 minutes to seconds.

$$8 \text{ minutes} = 8 \times 60 = 480 \text{ seconds}$$

Then multiply the speed of light by the number of seconds to get the distance of the Sun from the Earth.

$$480 \times 299\,792 = 144\,000\,000 = 1.44 \times 10^8 \text{ km (3 sf)}$$

Example 11

Use a calculator to work $(3.74 \times 10^4) \times (2.49 \times 10^5)$. Give your answer in standard form, correct to three significant figures (3 sf).

EXP **×10ˣ**

The key that you use on your calculator, to enter standard form, may be **×10ˣ** or **EXP**

To enter 3.74×10^4 press these keys.

3 **.** **7** **4** **EXP** **4**

Note that you do not press the × sign. If you press this key you will enter the wrong value.

To enter the full calculation, press these keys.

3 **.** **7** **4** **EXP** **4** **×** **2** **.** **4** **9** **EXP** **5** **=**

The display will be 9312600000 or 9.3126×10^9.

When rounded to three significant figures, the answer is 9.31×10^9.

Ensure that you are familiar with the way your calculator shows standard form.

Exercise 8E

  1 Complete each calculation. Give your answers in standard form.

a $(2 \times 10^3) \times (4 \times 10^2)$ b $(3 \times 10^2) \times (4 \times 10^5)$ c $(4 \times 10^3) \times (2 \times 10^4)$

d $(3 \times 10^5) \times (3 \times 10^8)$ e $(4 \times 10^5) \times (8 \times 10^{12})$ f $(6 \times 10^3) \times (7 \times 10^{10})$

g $(4.2 \times 10^4) \times (5 \times 10^2)$ h $(2.5 \times 10^4) \times (4 \times 10^5)$ i $(7 \times 10^6) \times (8 \times 10^4)$

j $(2.5 \times 10^6) \times (9 \times 10^3)$ k $(2.8 \times 10^5) \times (4 \times 10^6)$ l $(6 \times 10^3)^2$

 2 Complete each calculation. Give your answers in standard form. Do not round off your answers.

a $(4.3 \times 10^4) \times (2.2 \times 10^5)$ b $(6.4 \times 10^2) \times (1.8 \times 10^5)$

c $(2.8 \times 10^2) \times (4.6 \times 10^7)$ d $(1.9 \times 10^3) \times (2.9 \times 10^5)$

e $(7.3 \times 10^2) \times (6.4 \times 10^6)$ f $(9.3 \times 10^4) \times (1.8 \times 10^6)$

g $(3.25 \times 10^4) \times (9.2 \times 10^1)$ h $(2.85 \times 10^4) \times (4.6 \times 10^{12})$

i $(3.6 \times 10^2)^2$ j $(8.1 \times 10^5)^2$

 3 Complete each calculation. Give your answers in standard form. Round your answers to three significant figures.

a $(2.35 \times 10^5) \times (4.18 \times 10^5)$ b $(1.78 \times 10^5) \times (4.09 \times 10^2)$

c $(9.821 \times 10^2) \times (7.402 \times 10^6)$ d $(2.64 \times 10^2) \times (8.905 \times 10^5)$

e $(4.922 \times 10^4) \times (8.23 \times 10^8)$ f $(7.92 \times 10^3) \times (7.38 \times 10^6)$

g $(4.27 \times 10^6) \times (6.92 \times 10^8)$ h $(2.65 \times 10^5) \times (5.87 \times 10^{12})$

i $(7.83 \times 10^6)^2$ j $(2.534 \times 10^4)^2$

4 USB memory sticks are used to store large amounts of information.

One gigabyte (GB) is a billion bytes.

How many bytes can be stored on a 256 GB USB memory stick?

Give your answer in standard form.

 5 At the end of a hot summer, the contents of a reservoir were reduced to only 1.8×10^6 litres of water.

Then the rain came and water from rivers and streams flowed into the reservoir, so the increase of water in the reservoir was estimated at 1.2×10^5 litres per day.

It rained at the same rate for 40 days.

After 40 days, how much water was there in the reservoir?

Challenge: Mega-memory

The table shows the words and sizes of memory in the computer world.

1000	KB	kilobyte
1000^2	MB	megabyte
1000^3	GB	gigabyte
1000^4	TB	terabyte
1000^5	PB	petabyte
1000^6	EB	exabyte
1000^7	ZB	zettabyte
1000^8	YB	yottabyte

A Rewrite the table, putting the numbers in the first column in standard form.

B How many bytes of memory could there be in a database with a capacity of a million terabyte?

C A database was built with 5 million yottabyte of memory.

How much is this? Give your answer in standard form.

Ready to progress?

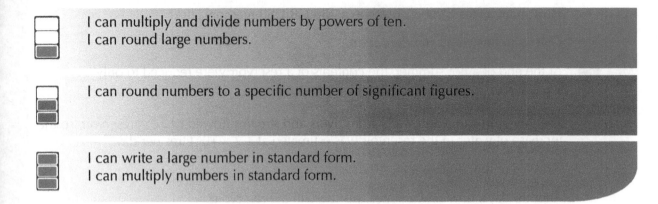

I can multiply and divide numbers by powers of ten.
I can round large numbers.

I can round numbers to a specific number of significant figures.

I can write a large number in standard form.
I can multiply numbers in standard form.

Review questions

1   What is the volume of a cube of side 1000 mm?

2   Joe was on a salary of £16 500. After a promotion he was awarded a 27% pay rise.
    What is Jo's salary now? Give your answer correct to two significant figures.

3   At an outdoor concert there was a huge crowd.
    Arran, Billy and their dad, Dean, were all at the concert.

It looks to me as though there are about seventy thousand fans here tonight.

I'm sure I heard them announce there are under seventy thousand here tonight.

I believe you're both correct.

    Explain how they could both have been correct.

4   Look at this sequence.
    one, one hundred, ten thousand, one million...
    Write down the next three terms.

5   a   Express:

      i   0.435 as a percentage

      ii   4 350 000 in standard form

      iii   4.3553 correct to three significant figures.

   b   Write these numbers in order of size, starting with the smallest.

      5 million 4.3×10^7 899 999 2250^2

6   What is the square root of 2.56×10^{24}?

7   In 2013 one hundred and fifty thousand people were expected to attend the Glastonbury Rock Festival.

   a   If these numbers were correct to two significant figures, what were the smallest and greatest numbers of people expected to attend?

      In 2000 there were sixty thousand cars at the festival.

      In 2007 there were thirty thousand cars at the festival.

   b   If this rate continued, what would you expect the greatest number of cars to be at the festival in 2014? Explain your answer.

MR   8   The nearest star to the Earth (apart from the Sun) is Proxima Centauri. It is 4.22 light years away. One light year is 9.46×10^{12} km.

   a   What is the distance, in kilometres, from the Earth to Proxima Centauri?

   b   Buzz said: 'A spacecraft travelling at 40 000 kilometres per hour would take about 114 thousand years to get there.'

      Show that Buzz is correct.

MR   9   Read what Professor Higgins is saying.

I don't like rounding to one significant figure, I could be as much as a third out.

Show that Professor Higgins is correct.

Challenge

Space – to see where no one has seen before

It is possible to see objects in space that are a staggering 13.8 billion light years away. A light year is the distance that light travels in one Earth year. Because it takes light so long to reach us from distant objects, we do not even know whether they still exist by the time we see them!

Although we have not yet seen anything that is further away than 13.8 billion light years we think that the diameter of the universe, within which we can observe objects – if they exist – is 92.876 billion light years. We think of this observable universe as a sphere with the observers (us) at the centre, so its edge is half of the diameter away from us, that is, 46.438 billion light years.

1 Light travels at 299 792 458 metres per second. Find, in standard form, the distance light will travel in a year (a light year).

2 Find the furthest distance, in kilometres written in standard form, that we can observe in space.

3 Rewrite the information in these questions on this page, correcting each number to one significant figure.

Professor Ball estimated that there are 152 317 298 405 stars in our galaxy.

He also estimated that there are 163 724 308 912 galaxies in the universe.

4 Give the number of stars in our galaxy in standard form, correct to three significant figures.

5 Give the number of galaxies in the universe in standard form, correct to three significant figures.

6 If we assume all galaxies have a similar number of stars, then how many stars are there altogether in the universe? Give your answer in standard form, correct to two significant figures.

The observable universe is a sphere with a diameter of 8.871×10^{26} m. The formula for finding the volume of a sphere is $V = 4(3.14 \times radius^3) \div 3$.

7 Find the volume of a sphere with a radius of 8 cm.

8 Find the volume of a sphere with a diameter of 8 cm.

9 Find the volume of the observable universe. Give your answer in standard form, correct to two significant figures.

9

Interpreting data

Distribution of the securities market key players

This chapter is going to show you:

- how to interpret pie charts by the angle size of each sector
- how to use the scaling method to construct pie charts
- how to use scatter graphs
- how to construct scatter graphs.

You should already know:

- how to interpret data from tables, graphs and charts
- how to read pie charts where the sectors are percentages or simple fractions
- how to find the mode, median, mean and range for small data sets.

About this chapter

Understanding data can help you to see what to do, both now and in the future, and sometimes what not to do! You will need to make sense of the huge amount of data presented to you on TV and via the internet, newspapers and magazines. Then you should look at the conclusions other people draw from data and decide whether you agree with them.

The way that data is represented can make it misleading. If you know how to spot this, it will help you to understand how to present data yourself, to show conclusions or to make a point clearly.

In this chapter you will look at some commonly used types of statistical diagrams – pie charts, line graphs and scatter graphs – and learn how to interpret them correctly and create them yourself.

9.1 Pie charts

Learning objective

- To work out the size of sectors in pie charts by their angles at the centre

Key words

| pie chart | sector |

You already know that **pie charts** can show proportions in a set of data, either as percentages or fractions.

For example, this pie chart shows the proportion of British and foreign cars sold over one weekend at a car salesroom.

The salesroom sold a total of 40 cars.

You can see from the pie chart that $\frac{1}{4}$ of the cars sold were British and $\frac{3}{4}$ were foreign. So:

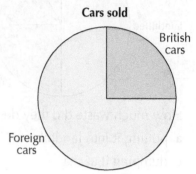

Cars sold

British cars

Foreign cars

- $\frac{1}{4}$ of 40 = 10   So 10 British cars were sold.

- $\frac{3}{4}$ of 40 = 30   So 30 foreign cars were sold.

How do you read a pie chart when it is divided into fractions that are not so obvious?

You need to work out the fraction of the whole pie chart that is taken up by each **sector**.

You can do this by:

- working out the angle the sector makes at the centre of the pie chart

- calculating this angle as a fraction of the total angle of 360°.

This will tell you the fraction of the data that is represented by the sector.

Example 1

The pie chart shows the types of housing on a new estate.

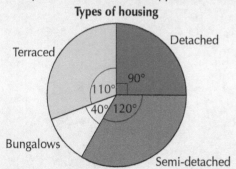

Types of housing

Detached

Terraced

110° 90°

40° 120°

Bungalows

Semi-detached

Altogether there were 540 new houses built.

How many were:

a detached   **b** semi-detached   **c** bungalows   **d** terraced?

You need to work out the fraction of 540 that each sector represents.

a $\frac{90}{360} \times 540 = 135$ detached   **b** $\frac{120}{360} \times 540 = 180$ semi-detached

c $\frac{40}{360} \times 540 = 60$ bungalows   **d** $\frac{110}{360} \times 540 = 165$ terraced

Example 2

The pie chart shows how one country dealt with 3000 kg of dangerous waste in 2013.

Disposing of dangerous waste

How much waste did they destroy by:

a putting it into landfill   **b** burning it

c dumping it at sea   **d** chemical treatment?

From the pie chart you can easily see that:

a $\frac{1}{2}$ of the waste was put into landfill.

$\frac{1}{2}$ of 3000 kg = 1500 kg

So 1500 kg of the waste was put into landfill.

b $\frac{1}{4}$ of the waste was burnt.

$\frac{1}{4}$ of 3000 = 3000 ÷ 4 = 750

So 750 kg of the waste was burnt.

c The angle of this sector is 30°.

Then the fraction of the circle is:

$$\frac{30}{360} = \frac{1}{12}$$

Therefore $\frac{1}{12}$ of the waste was dumped at sea.

3000 ÷ 12 = 250

So 250 kg of the waste was dumped at sea.

d The angle of this sector is 60°.

Then the fraction of the circle is:

$$\frac{60}{360} = \frac{1}{6}$$

Therefore $\frac{1}{6}$ of the waste was treated by chemicals.

3000 ÷ 6 = 500

So 500 kg of the waste was treated by chemicals.

Exercise 9A

1 All 900 pupils in a school were asked to vote for their favourite subject.

The pie chart illustrates their responses.

How many voted for:

a PE **b** science **c** mathematics?

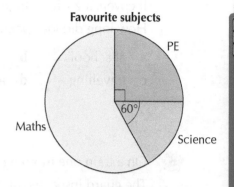

Favourite subjects

2 One weekend in Edale, a café sold 300 drinks.

The pie chart illustrates the proportions of different drinks that were sold.

Of the drinks sold that weekend, how many were:

a soft drinks **b** tea
c hot chocolate **d** coffee?

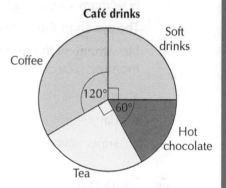

Café drinks

3 In one week, a supermarket sold 540 kilograms of butter.

The pie chart shows the proportions that were sold on different days.

How many kilograms of butter were sold on:

a Monday **b** Tuesday **c** Wednesday
d Thursday **e** Friday **f** Saturday
g Sunday?

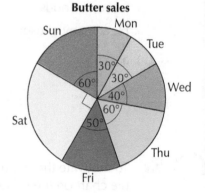

Butter sales

4 Priya did a survey about fruit and nut chocolate. She asked 30 of her friends.

The pie chart illustrates her results.

How many of Priya's friends:

a never ate fruit and nut chocolate

b sometimes ate fruit and nut chocolate

c said that fruit and nut was their favourite chocolate?

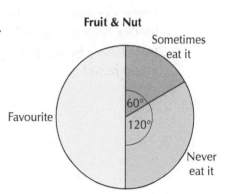

Fruit & Nut

5 The pie chart shows the daily activities of Joe one Wednesday.

It covers a 24-hour time period.

How long did Joe spend:

a at school   b at leisure

c travelling   d asleep?

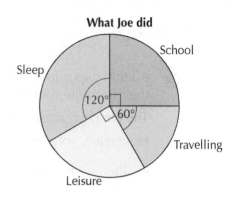

What Joe did

6 On a train one morning, there were 240 passengers.

The guard inspected all the tickets and reported how many of each sort there were.

The pie chart illustrates his results.

How many of the tickets that he saw that morning were:

a open returns   b season tickets

c day returns   d travel passes

e super savers?

Train tickets

7 The pie chart shows the results of a survey of 216 children, when they were asked about their favourite foods.

How many chose:

a chips   b pizza

c pasta   d curry?

Favourite food

(PS) **8** To motivate the pupils, a headteacher placed this pie chart on the school noticeboard.

Emma decided to use the chart to find out how much money each year group had raised.

Help Emma by estimating how much each year group raised.

Christmas charity collection
total so far … £1690

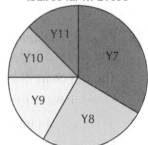

PS **9** Brendan saw this pie chart in a magazine.

Brendan measured and recorded the angles of the pie chart in a table, like this.

Northern Ireland	11°
Wales	53°
Scotland	58°
England	238°

Calculate how many megalitres of water there were in the reservoirs of each country.

Give your answers to two significant figures.

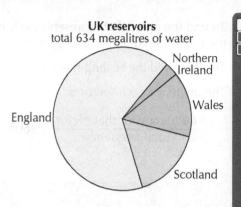

UK reservoirs
total 634 megalitres of water

Northern Ireland

Wales

England

Scotland

Activity: Population in a pie chart

Liam was looking through some statistics about the UK.

He found this unfinished pie chart, which shows the population by age group.

He was curious to know what the missing data was, and to find the actual numbers of people in each age group in the chart.

A State what the missing data is most likely to be.

B Find the actual population numbers for each age group.

C This was for the year 1989.

Do some research to find the current figures.

Would the pie chart be very similar?

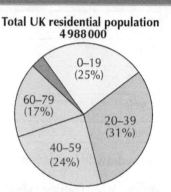

Total UK residential population
4 988 000

0–19 (25%)
60–79 (17%)
20–39 (31%)
40–59 (24%)

9.2 Creating pie charts

Learning objective

• To use a scaling method to draw pie charts

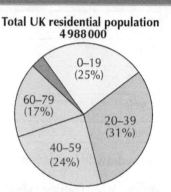

Key words	
frequency	frequency table
scaling	total frequency

Sometimes you will have to draw a pie chart to display data that is given in a **frequency table**.

Suppose you were asked to draw a pie chart to represent this set of data, showing how a group of people travel to work.

Type of travel	Walk	Car	Bus	Train	Cycle
Frequency	24	84	52	48	32

Each type of travel (walk, car, bus and so on) will be represented by its own sector in the pie chart.

The size of the sector will depend on the **frequency** for that sector.

The **total frequency** is 240 people.

A pie chart has an angle of 360° at its centre.

To find the angle that represents each frequency, you need to work out the frequency as a fraction of the total, then multiply that by 360°.

This is called the **scaling** method.

The angle for each sector is:

$$\frac{\text{frequency for that sector}}{\text{total frequency}} \times 360°$$

Example 3

Draw a pie chart to represent the data showing how a group of people travel to work.

Set the data out in a vertical table and write the calculations in it.

Sector (type of travel)	Frequency	Calculation	Angle
Walk	24	$\frac{24}{240} \times 360° = 36°$	36°
Car	84	$\frac{84}{240} \times 360° = 126°$	126°
Bus	52	$\frac{52}{240} \times 360° = 78°$	78°
Train	48	$\frac{48}{240} \times 360° = 72°$	72°
Cycle	32	$\frac{32}{240} \times 360° = 48°$	48°
Total	240		360°

Now draw the pie chart.

 Hint When drawing a pie chart, draw the smallest angle first and try to make the largest angle the last one you draw, then any cumulative error in drawing will not be so noticeable.

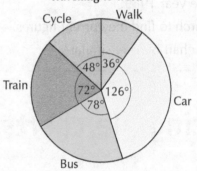

Exercise 9B

1 Draw a pie chart to represent the favourite subjects of 36 pupils.

Subject	Maths	English	Science	Languages	Other
Frequency	12	7	8	4	5

2 Draw a pie chart to represent the types of food that 40 people usually eat for breakfast.

Food	Cereal	Toast	Fruit	Cooked	Other	None
Frequency	11	8	6	9	2	4

3 Draw a pie chart to represent the numbers of goals scored by an ice-hockey team in 24 matches.

Goals	0	1	2	3	4	5 or more
Frequency	3	4	7	5	4	1

4 Draw a pie chart to represent the favourite colours of 60 Year 8 pupils.

Colour	Red	Green	Blue	Yellow	Other
Frequency	17	8	21	3	11

PS **5** This is an estate agent's waiting list for rental flats on one day in a large city.

Type of flat	Number of applicants
1 bedroom	646
2 bedrooms	1344
3 bedrooms	80
4 bedrooms	6
5 bedrooms	2

    a Why would it be extremely difficult to draw a pie chart to illustrate this information?

    b Combine the groups so that you could draw a pie chart to represent this information.

    c Draw the pie chart.

PS **6** This diagram was shown in a business magazine.

It illustrates the funds received by a charity one year.

Redraw it as an accurate pie chart.

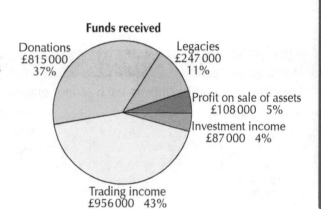

Funds received

Donations £815 000 37%
Legacies £247 000 11%
Profit on sale of assets £108 000 5%
Investment income £87 000 4%
Trading income £956 000 43%

Challenge: World energy consumption

The bar chart illustrates the world energy consumption, given as percentages of the total energy used.

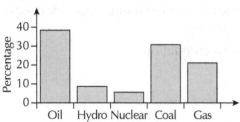

A Find the percentage, as accurately as you can, for each type of energy.

B Explain why your total may not come to 100%.

C Draw a pie chart to represent this data.

9.3 Scatter graphs and correlation

Learning objectives

- To read scatter graphs
- To understand correlation

Key words

| correlation | scatter graph |

A doctor recorded the brightness of the sunlight and the size of the pupils of people's eyes.

She then plotted the results on a graph. The result is a scattering of points on the graph, which is why this type of graph is called a **scatter graph**.

What can you tell about the connection between the brightness of the sunlight and the size of people's pupils?

You can see that as the light gets brighter the size of people's pupils gets smaller.

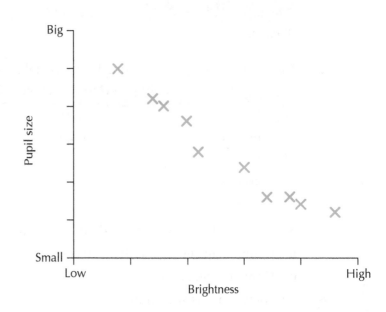

Example 4

Describe the relationships in each scatter graph.

a

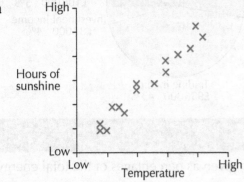

This graph shows that the temperature is higher when there are more hours of sunshine.

This is a positive **correlation**.

b

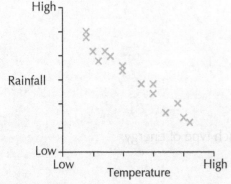

This graph shows that as the temperature increases, the rainfall decreases.

This is a negative correlation.

c

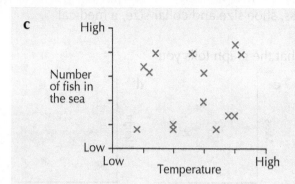

This graph shows no correlation. This means that there is no connection between the temperature and the number of fish in the sea.

In summary:

- positive correlation means that the bigger one factor is, the bigger the other is as well
- negative correlation means that the bigger one factor is, the smaller the other one is
- no correlation means that there is no connection between the two sets of data.

Note that a simple correlation does not tell you why there is a connection between two sets of data. Also, one factor might not affect the other. You could have a graph showing that sales of ice cream and sun cream have positive correlation, but the cause of both is a third factor – summer sunshine!

Exercise 9C

 1 After completing a study into temperature, rainfall and the sales of ice creams, umbrellas and the hire of deckchairs, a tour guide drew these scatter graphs. Describe each type of correlation and explain why there might be that correlation.

a

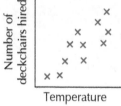

b

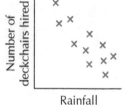

c

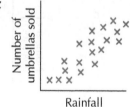

d

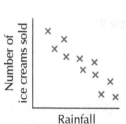

e

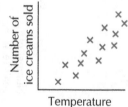

f

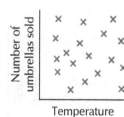

2   After completing a study into height, mass, shoe size and collar size, a medical student drew these scatter graphs.

Describe each type of correlation and what the graph tells you.

a   **b**   **c**   **d**

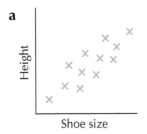   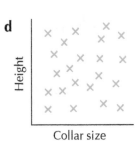

3   After a series of mathematics, science and English test results, a teacher drew these scatter graphs.

Describe the type of correlation and what each graph tells you.

a   **b**   **c**   **d**

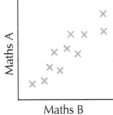  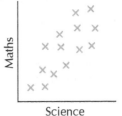 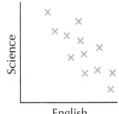

  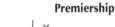

MR   4   A fan surveyed the transfer price and age of some goalkeepers, as well as how many goals they let in during their first full season. She drew these scatter graphs.

Describe the type of correlation and why there might be that correlation.

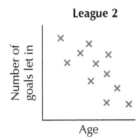

MR **5** Pat studied the mass, distance to travel, price of postage and how long it took for parcels to be delivered by Royal Mail. He drew these scatter graphs.

Describe the type of correlation and what each graph tells you.

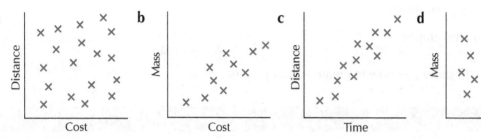

PS **6** These three scatter graphs show the heights of people and the cost of their clothes.

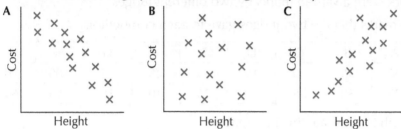

Billy said: 'The shorter you are, the less your clothes will cost.'

Terry said: 'The shorter you are, the more your clothes will cost.'

Suzie said: 'Your height doesn't make any difference to the cost of your clothes.'

Match up the correct graph with the statements each person has made.

7 Sketch a possible scatter graph to illustrate each statement.

a The bigger the bird, the higher it can fly.

b The cheaper the phone, the less likely it is to get stolen.

c The hotter the weather, the fewer thick coats are sold.

d There is no correlation between the temperature and how many newspapers are sold.

PS **8** 'There was a time when people would have to go to the cinema to see films. Then video recorders became popular and fewer people went to cinemas, so many of them shut down. Now everyone has DVD players but there has been a huge return to the cinema.'

There are three stages of development here. Scatter graphs can help to portray each one. Sketch three scatter graphs to help illustrate this story.

Activity: Correlation in circles

Try to find out if there is any correlation between the diameter of a circle and its circumference.

• Use a variety of circular objects, such as tins and coins.

• Use compasses to draw some circles.

• Measure the diameter and the circumference of each circle and record your measurements.

Is there a pattern?

9.4 Creating scatter graphs

Learning objective

• To create scatter graphs

This section will show you how to create scatter graphs.

Example 5

Ten people entered a craft competition.

Their displays of work were awarded marks by two different judges.

The table shows the marks that the two judges gave to each competitor.

Competitor	A	B	C	D	E	F	G	H	I	J
Judge 1	90	35	60	15	95	25	5	100	70	45
Judge 2	75	30	55	20	75	30	10	85	65	40

a Draw a scatter graph to illustrate this information.

b Is there any correlation between the two judges and, if so, what type?

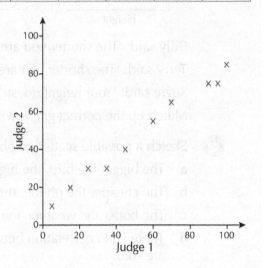

a Use a scale of 0 to 100 on each axis.

Label the horizontal axis as Judge 1 and the vertical axis as Judge 2.

For each contestant, use the two marks to form a pair of coordinates, (Judge 1, Judge 2).

Plot the points on the grid.

The scatter graph will look like this.

b There is a positive correlation between the judges.

Exercise 9D

1 Dan carried out a survey to compare the ages of some people in his school with the amount of money that they spend each week.

Age (years)	11	16	14	13	13	16	10	12	14	15
Amount spent (£)	5	6	8	7	9	14	5	6	10	11

a Plot the data on a scatter graph.

Label the horizontal axis as 'Age (years)', from 10 to 16.

Label the vertical axis as 'Amount spent (£)', from 0 to 15.

b Describe in words what the graph tells you and what sort of correlation there is.

c Use your graph to estimate what age a pupil might be if they spent £7.70 each week.

2 The table shows how much time pupils spend watching television and how long they spend on homework per week.

Time watching TV (hours)	12	8	5	7	9	3	5	6	10	14
Time spent on homework (hours)	4	7	10	6	5	11	9	6	6	3

a Plot the data on a scatter graph. Label each axis as 'Time (hours)', from 0 to 15.

b Describe in words what the graph tells you and what sort of correlation there is.

c Use your graph to estimate how much time a pupil spent watching TV in the week if they spent 4 hours on homework in the week.

3 The table shows the price paid for some cars of different ages.

Age (years)	1	2	3	4	5	6	7
Amount spent (£)	15 000	13 000	10 500	8000	6300	4200	3400

a Plot the data on a scatter graph. Label the horizontal axis as 'Age (years)', from 1 to 10.

Label the vertical axis as 'Amount spent (£)', from 1000 to 15 000.

b Explain what the graph tells you about what happens to a car's value as it gets older.

c Estimate how much the car was worth when it was three and a half years old.

PS **4** A teacher had given her class two mathematics tests, but four pupils were absent for one or the other of them. These are the class results.

Pupil	Test A	Test B
Andy	69	59
Ben	34	17
Celia	8	abs
Dot	42	43
Eve	77	54
Faye	72	43
Gill	54	38
Harry	40	25
Ida	64	abs

Pupil	Test A	Test B
Joy	76	61
Kath	85	65
Les	abs	41
Meg	35	30
Ned	14	15
Olly	82	63
Pete	36	35
Quale	20	30
Jess	63	42

Pupil	Test A	Test B
Robin	48	32
Sophia	71	52
Tom	52	49
Ulla	abs	28
Vera	81	62
Will	41	36
Xanda	58	48
Yin	40	32
Zeb	28	19

Use the test scores to create a scatter graph and so estimate what score the absent pupils might have been expected to get if they had taken the missing test.

Problem solving: Fish food

In the sea around Statsland, the numbers of fish and whales follow a cycle.

- The fish increase in number, which attracts the whales.
- The whales eat the fish and grow in number, eating more fish.
- This leads to fewer fish, so the whales move elsewhere.
- This allows the fish population to grow again, which then attracts the whales – and the cycle starts all over again.

Draw four separate scatter graphs to illustrate this cycle of fish and whales.

Ready to progress?

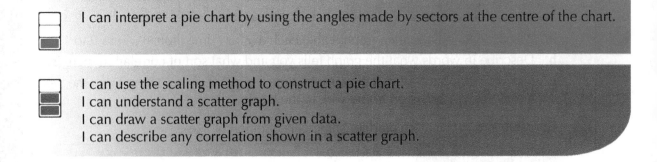

I can interpret a pie chart by using the angles made by sectors at the centre of the chart.

I can use the scaling method to construct a pie chart.
I can understand a scatter graph.
I can draw a scatter graph from given data.
I can describe any correlation shown in a scatter graph.

Review questions

1   There are 120 pupils in a year group. 12 of these pupils wear glasses.

    a   The pie chart to show this is not drawn accurately.
       What should the angles be? Show your working.

    b   Exactly half of the 120 pupils in the school are boys.

       From this information, is the percentage of boys in this school that wear glasses 5%, 6%, 10%, 20%, 50% or is it not possible to tell?

Who wears glasses

Wear glasses

Do not wear glasses

2   A teacher asked two different classes: 'What is your favourite type of book?'

    a   The table shows the results from class A (total 30 pupils).

Type of book	Frequency
Crime	5
Non-fiction	15
Fantasy	10

       Draw a pie chart to show this information. Show your working and draw your angles accurately.

    b   The pie chart shows the results from all of class B.
       Each pupil had only one vote.
       The sector for non-fiction represents 11 pupils.
       How many pupils are there in class B?
       Show your working.

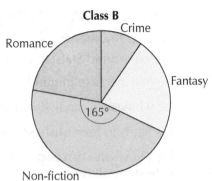

Class B

Crime

Romance

Fantasy

165°

Non-fiction

3   The scatter graph shows 15 pupils' marks in mathematics and English tests.

To find a pupil's total mark, you add the mathematics mark to the English mark.

a   Which pupil had the highest total mark?

b   Is the statement below true or false?

'The range of English marks was greater than the range of mathematics marks.'

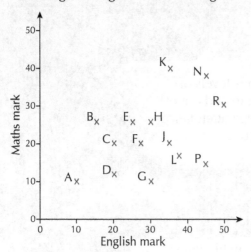

c   Pupils with total marks in the shaded region on the graph below win a prize.
What is the smallest total mark needed to win a prize?

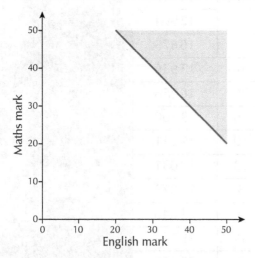

PS 4   Andrew thought his coffee always cooled down quickly. One day, he recorded the temperature of his coffee at various times after he had made it. His results are in this table.

Time (minutes)	0	1	2	3	4	5	6	7	8	9	10	12	14	16	18	20
Temperature (°C)	100	95	80	73	65	58	52	45	40	36	30	22	20	18	17	16

a   Draw a scatter graph from the results.

b   What type of correlation is there between the time and the temperature over the first ten minutes?

c   Andrew's Grandpa said that the perfect 'drinking' temperature for coffee is 70 °C.
Andrew decided to try out his Grandpa's suggestion. How long should Andrew wait, after making his coffee, before drinking it?

Challenge
Football attendances

The Sheffield football teams have had various levels of support over the years, as they have moved in and out of the divisions. This table compares the average attendances of their supporters over the years 1995 to 2013.

Year	Sheffield Wednesday			Sheffield United		
	Level	Place	Attendance	Level	Place	Attendance
1995	1	13	26 596	2	8	14 408
1996	1	15	24 877	2	9	12 904
1997	1	7	25 714	2	5	16 675
1998	1	13	28 706	2	6	17 936
1999	1	12	26 745	2	8	16 258
2000	1	19	24 855	2	16	13 700
2001	2	17	19 268	2	10	17 211
2002	2	20	20 882	2	13	18 031
2003	2	21	20 327	2	3	18 073
2004	3	16	22 336	2	8	21 646
2005	3	5	23 100	2	8	19 594
2006	2	19	24 853	2	2	23 650
2007	2	9	23 638	1	18	30 512
2008	2	16	21 418	2	12	25 631
2009	2	12	21 541	2	15	26 023
2010	2	22	23 179	2	13	25 120
2011	3	15	17 817	2	17	20 632
2012	3	2	21 336	2	21	18 701
2013	2	18	23 413	3	3	18 611

1 Draw a pie chart for each team, showing how many years they have spent in the different levels.

2 Put the data into a grouped bar chart, showing the number of times the average attendances were in the bands 10 000–15 000, 15 001–20 000, 20 001–25 000, 25 001–30 000, over 30 000. Comment on the differences in attendance between the teams.

3 What is the range of the average attendances for each club?

4 What is the mean level played at by each team during these years?

5 Draw a scatter graph for each team, using attendance and place, assuming that Level 1 has 20 places, Level 2 has 22 places and Level 3 has 22. Comment on anything these show about attendance and place for each team.

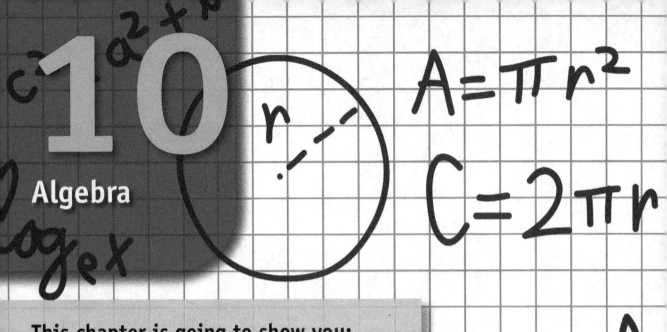

10

Algebra

This chapter is going to show you:

- how to write and simplify expressions involving all four operations
- how to simplify expressions that have a number of terms
- how to multiply out brackets in algebraic expressions
- how to identify equivalent expressions
- how to write algebraic expression involving powers.

You should already know:

- what an algebraic expression is
- how to interpret simple algebraic expressions
- how to write simple algebraic expressions.

About this chapter

Spreadsheets are used in many areas of business and industry. They can be used to carry out any sort of calculation you require. A common example is financial calculations involving costs, quantities, profits, tax or interest. The calculations needed are written in symbolic form in the spreadsheet. This symbolic form is what we mean by algebra.

Algebra is very useful in science, and in areas such as computer programming, because it helps us to write things simply and clearly. The rules of algebra are like the grammar of a language. If you know them, you can read algebraic sentences and write them yourself. This chapter is about some of the rules of algebra.

10.1 Algebraic notation

Learning objective

- To simplify algebraic expressions involving the four basic operations

Key word

algebraic expression

You already know that you usually leave out the multiplication sign (×) when you write **algebraic expressions**.

For example, instead of $4 \times a$ or $a \times 4$, you should write $4a$.

Notice that you always put a number in front of a letter. Never write it as $a4$. You will see why this is important, in section 10.5.

You do not usually use the division sign (÷) either.

$a \div 4$ is written as $\frac{a}{4}$.

You can also use a fraction to write $a \div 4$ as $\frac{1}{4}a$. This is because to find a quarter of a number you divide it by 4. The two operations give the same answer.

Example 1

Write each of these as simply as possible.

a $3 \times t$ **b** $k \times 2 + 4$ **c** $d \times (8 - 2.5)$

  a $3 \times t = 3t$ You can leave out the × sign.

  b $k \times 2 + 4 = 2k + 4$ Put the 2 in front of the k.

  c $d \times (8 - 2.5) = d \times 5.5$ Work out the subtraction in the brackets first.

      $d \times 5.5 = 5.5d$ Put the 5.5 in front of the d.

If you have more than two things to multiply, you can do them in any order.

If a letter is multiplied by itself, use the 'square' notation to write it down.

For example, write $d \times d$ as d^2.

Example 2

Write each of these as simply as possible.

a $p \times k \times r$ **b** $2 \times k \times 4$ **c** $f \times \frac{2}{3} \times f$

  a $p \times k \times r = pkr$ You do not have to put the letters in alphabetical order.

  b $2 \times k \times 4 = 8k$ Multiply the numbers first.

  c $f \times \frac{2}{3} \times f = \frac{2}{3}f^2$ Put the $\frac{2}{3}$ in front and then $f \times f = f^2$.

1 Write these expressions as simply as possible.

   a $4 \times k$ **b** $t \times 3$ **c** $2 \times x$ **d** $y \times 20$

   e $a \times b$ **f** $n \times m$ **g** $e \times \frac{1}{2}$ **h** $r \times \frac{1}{4}$

2 Write these expressions as simply as possible.

   a $2 \times a \times b$ **b** $4 \times 2 \times 5$ **c** $2 \times 5 \times e$ **d** $4 \times 20 \times w$

   e $a \times b \times c$ **f** $n \times 6 \times \frac{1}{2}$ **g** $2.5 \times x \times 3$ **h** $b \times \frac{3}{4} \times h$

3 Write these expressions as simply as possible.

   a $a \times a$ **b** $4 \times x \times x$ **c** $9 \times n \times n$ **d** $t \times t \times 1.4$

   e $2 \times a \times 1.6$ **f** $d \times \frac{1}{5} \times d$ **g** $\frac{1}{2} \times x \times x$ **h** $\frac{2}{3} \times m \times m$

4 Write these expressions without using a $\times$ sign.

   a $2 \times (n + 1)$ **b** $4 \times (t + 12)$ **c** $8 \times (3 + k)$ **d** $0.5 \times (t - 6.4)$

   e $(k + 2) \times 4$ **f** $(12 - y) \times 2$ **g** $(2 + 6) \times (x - 2)$ **h** $(40 - y) \times 3 \times 2$

5 Write each expression as simply as possible.

   a $2 + 3 \times a$ **b** $(5 + 4) \times x$ **c** $w \times (3.5 + 4.2)$ **d** $a \times b - 1.4$

   e $2 + 3 + 4 \times a$ **f** $d \times (10 - 1.5)$ **g** $12 - 7 \times n$ **h** $f \times 13 + 9$

6 Write each expression as simply as possible.

   a $2 \times 3n$ **b** $4 \times 5b$ **c** $4 \times 0.5d$ **d** $0.1 \times 5q$

   e $5 \times 2k$ **f** $4g \times 6$ **g** $8t \times 1.5$ **h** $0.2 \times 5h$

7 Write each of these as simply as possible.

   a $2n \times 2n$ **b** $3d \times 4d$ **c** $4p \times 2p$ **d** $0.5a \times 6a$

8 Write each expression without using a division ($\div$) sign. The first one has been done for you.

   a $x \div 3 = \frac{x}{3}$ **b** $y \div 4$ **c** $t \div 1.5$ **d** $24 \div n$

9 In each expression, an algebraic unknown is divided by a whole number. Write the expression by using a fraction. The first one has been done for you.

   a $h \div 5 = \frac{1}{5}h$ **b** $m \div 3$ **c** $n \div 4$ **d** $ab \div 2$

(PS) 10 Match each expression on the top row with an equivalent expression on the bottom row.

   a $2 \times x + 3 \times y$ **b** $y \times (2 + 3) \times x$ **c** $3y \times 2x$ **d** $x + (2 + 3) \times y$

$x + 5y$	$6xy$	$3y + 2x$	$5xy$

Challenge: Is it true?

Some of these statements are always true, whatever values you choose for a and b.

i $a + b + 2 = 2 + a + b$ **ii** $a + 2 - b = a - 2 + b$

iii $a \times b + 2 = 2 + a \times b$ **iv** $\dfrac{a}{2} + b = a + \dfrac{b}{2}$

A Which ones are always true?

B Explain how you know that the others are not true.

10.2 Like terms

Learning objective

* To simplify algebraic expressions by combining like terms

Key words

| like terms | term |

Look at this expression:

$$5a + 4b - 3a + b$$

It has four **terms** altogether. There are two terms that include a and two terms with b.

You can write the terms in a different order like this:

$$5a - 3a + 4b + b$$

You can combine the a-terms and you can combine the b-terms.

$$5a - 3a = 2a \text{ and } 4b + b = 5b$$

So $5a - 3a + 4b + b = 2a + 5b$

$5a$ and $3a$ are called **like terms** and like terms can be combined to simplify an expression.

$4b$ and b are also like terms.

$2a$ and $5b$ are not like terms, because they contain different letters.

The expression $2a + 5b$ cannot be simplified any further.

Example 3

Simplify each expression as much as possible.

a $4f - 3f + 6$ **b** $2g - 3h + 2g - 4h$ **c** $x^2 - 4x + x - 5$

a $4f - 3f + 6 = f + 6$

The first two terms are similar and can be combined.

The 6 and the f are not similar terms and the expression cannot be simplified further.

b $2g - 3h + 2g - 4h = 2g + 2g - 3h - 4h$

$\qquad\qquad\qquad\quad = 4g - 7h$

Put like terms together.

Combine the g-terms and the h-terms.

Notice that subtracting $3h$ and then subtracting another $4h$ is the same as subtracting $7h$.

c $x^2 - 4x + x - 5 = x^2 - 3x - 5$

Only the middle terms are similar and can be combined.

Subtracting $4x$ and then adding one x is the same as subtracting $3x$.

1 Simplify each expression.

a	$5h + 6h$	**b**	$4p + p$	**c**	$9u - 3u$
d	$3b - b$	**e**	$-2j + 7j$	**f**	$6pr - pr$
g	$2k + k + 3k$	**h**	$9y^2 - y^2$	**i**	$7d^2 - 2d^2$
j	$10i + 3i - 6i$	**k**	$2b - 5b + 6b$	**l**	$12ab + 5ab$
m	$3xy + 6xy$	**n**	$4p^2 + 7p^2$	**o**	$15ab - 10ab$
p	$5a^2 + 2a^2 - 3a^2$	**q**	$4fg - 6fg + 8fg$		

2 Simplify each expression.

a	$6h + 2h + 5g$	**b**	$4g - 2g + 8m$	**c**	$8f + 7d + 3d$
d	$4x + 5y + 7x$	**e**	$6 + 3r - r$	**f**	$4 + 5s - 3s$
g	$c + 2c + 3$	**h**	$12b + 7 + 2b$	**i**	$7w - 7 + 7w$
j	$2bf + 4bf + 5g$	**k**	$7d + 5d^2 - 2d^2$	**l**	$6t^2 - 2t^2 + 5t$
m	$4s - 7s + 2t$	**n**	$5h - 2h^2 - 3h$	**o**	$4y - 2w - 7w$

3 Simplify each expression.

a	$9e + 4 + 7e + 2$	**b**	$10u - 4 + 9u - 2$	**c**	$b + 3b + 5d - 2d$
d	$4 + 5c + 3 + 2c$	**e**	$1 + 2g - 3g + 5$	**f**	$9h + 4 - 7h - 2$
g	$7p^2 + 8 - 6p^2 - 3$	**h**	$14j^2 - 5j + 3j + 9$	**i**	$4t^2 - 5t + 2t - 6$
j	$2 + 5t - 9t + 3t^2$	**k**	$5p - 2q - 7p + 3q$	**l**	$12 + 5e - 4 - 9e$

4 Match each expression on the top row with the correct simplified form on the bottom row.

a $2y - x - 5y + 7x$ **b** $x + 5x - y - x$ **c** $2x - 3y + 2y + 3x$ **d** $y + 5x - 4y + x$

$$\boxed{5x - y \qquad 6x - 3y}$$

Investigation: Four expressions

A Here are four expressions.

$2n + 1$ $2n + 3$ $2n + 5$ $2n + 7$

a Work out the value of each expression if $n = 10$.

b Work out an expression for the sum of the four expressions.

Write the answer as simply as possible.

c Add up the four values from part **a**.

d Substitute $n = 10$ in your answer to part **b**.

You should get the same answer as you did for part **c**.

B Repeat part **A** with these four expressions.

$3n - 3$ $3n - 1$ $3n + 1$ $3n + 3$

10.3 Expanding brackets

Learning objective

● To remove brackets from an expression

Key words

equivalent	expand
multiply out	

You have seen how brackets can be used in algebraic expressions. In this section you will learn how to remove brackets and write expressions in different ways.

Look at these multiplications.

$$3 \times (5 + 2) = 3 \times 7 = 21$$

$$3 \times 5 + 3 \times 2 = 15 + 6 = 21$$

This shows that:

$$3 \times (5 + 2) = 3 \times 5 + 3 \times 2$$

You get the same result if you multiply 5 and 2 by 3 separately and add the results.

A diagram may make this clearer.

×	5	+ 2
3	15	+6

$15 + 6 = 21$

This is a general result. This means that it is always true, even if you replace one or more of the numbers by letters.

Here are some more examples.

● $2(a + 3) = 2a + 6$ This means $2 \times a$ added to 2×3.

So $2(a + 3)$ and $2a + 6$ are **equivalent** expressions.

×	a	+ 3
2	2a	+6

● $4(c - d) = 4c - 4d$ This means $4 \times c$ take away $4 \times d$.

×	c	− d
4	4c	−4d

This is called **multiplying out** or **expanding** an expression with brackets.

Example 4

Write each expression without brackets, as simply as possible.

a $3(t - 5)$ **b** $f(f + 4)$ **c** $2(a + 3) + 3(a - 4)$

a $3(t - 5) = 3t - 15$ This means $3 \times t$ minus 3×5.

b $3(2f + 4) = 6f + 12$ $3 \times 2f = 6f$ and $3 \times 4 = 12$

c $2(a + 3) = 2a + 6$ Multiply out the brackets separately.

$3(a - 4) = 3a - 12$

So $2(a + 3) + 3(a - 4) = 2a + 6 + 3a - 12$ Put the two expressions together.

$= 2a + 3a + 6 - 12$ Put like terms together.

$= 5a - 6$ $2a + 3a = 5a$ and $6 - 12 = -6$

Exercise 10C

1 Expand the brackets.

  a $5(p + 2)$ **b** $4(m - 3)$ **c** $2(t + u)$ **d** $4(d + 2)$

  e $5(b + 5)$ **f** $6(j - 4)$ **g** $2(5 + f)$ **h** $10(1 - n)$

2 Expand the brackets.

  a $2(a + b)$ **b** $3(q - t)$ **c** $4(t + m)$ **d** $5(x - y)$

  e $2(f + g)$ **f** $4(1.5 - f)$ **g** $1.5(h + 10)$ **h** $4(3.5 - f)$

3 Expand and simplify each expression.

  a $3w + 2(w + 1)$ **b** $5(d + 2) - 2d$ **c** $4h + 5(h + 3)$

  d $2x + 4(3 + x)$ **e** $2(m - 3) - 7$ **f** $16 + 3(q - 4)$

4 Expand and simplify the following expressions.

  a $4(a + 1) + 2(a + 2)$ **b** $3(i + 4) + 5(i + 2)$

  c $3(p + 2) + 2(p - 1)$ **d** $5(d - 2) + 3(d + 1)$

  e $4(e + 2) + 2(e - 3)$ **f** $2(x - 2) + 6(x + 1)$

  g $5(m - 3) + 3(m + 4)$ **h** $4(u - 3) + 5(u - 2)$

5 Write these expressions without brackets

  a $2(2a + 3)$ **b** $3(2x - 5)$ **c** $4(3t + 5)$ **d** $10(5n - 3)$

  e $6(5 + 2a)$ **f** $4(2 - 3y)$ **g** $20(3 + 4r)$ **h** $5(7 - 2m)$

(MR) 6 Here are three expressions.

  $2(x + 3)$ $4(1 - x)$ $2(x - 5)$

  a Find the sum of the three expressions.

      Write your answer as simply as possible.

  b If $x = 7$, find the value of each of the three expressions and check that your answer to part **a** is correct.

Challenge: Equivalent expressions

Here are some expressions.

Divide them into groups so that all the expressions in a group are equivalent.

  a $4(x + 3)$ **b** $4(x + 2)$ **c** $2x + 2(x + 4)$

  d $2(x + 1) + 2(x + 5)$ **e** $x + 3(x + 4)$ **f** $2(2x + 4)$

  g $4 + 4(x + 1)$ **h** $2(2x + 6)$ **i** $3(x + 5) + x - 3$

10.4 Using algebraic expressions

Learning objectives

- To manipulate algebraic expressions
- To identify equivalent expressions

Key word

manipulate

In mathematics, and in other subjects, you can use algebraic expressions to show relationships.

Sometimes expressions can be written in different ways.

The shape below has been divided into two rectangles in two different ways.

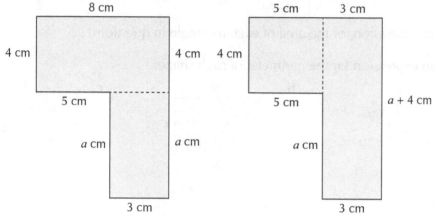

Some lengths are unknown, so they have been labelled with a letter.

The areas of the rectangles on the left are $4 \times 8 = 32$ and $3 \times a = 3a$.

Then the total area is $32 + 3a$.

The areas of the two rectangles on the right are $4 \times 5 = 20$ and $3(a + 4)$.

So the total area is $20 + 3(a + 4)$.

This means that $32 + 3a$ and $20 + 3(a + 4)$ must be equivalent.

You can show this is true by multiplying out the term with the brackets.

$20 + 3(a + 4) = 20 + 3a + 12$ because $3(a + 4) = 3a + 12$.

When you use algebraic expressions in this way you are **manipulating** them.

Example 5

In this diagram, all the lengths are in centimetres.

Show that the perimeter of this shape is $6(x + 6)$.

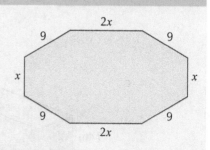

The perimeter is
$2x + 9 + x + 9 + 2x + 9 + x + 9$.   Add the lengths of the 8 sides.

This is $2x + x + 2x + x + 4 \times 9$ Four of them are 9 cm.

$= 6x + 36$ Add like terms.

This is the same as the expanded expression so the perimeter of the shape is $6(x + 6)$.

The expression is correct.

1   Work out an expression for the perimeter of each rectangle.
Write your answer as simply as possible.

a 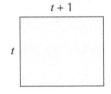   **b** 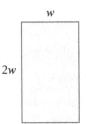   **c**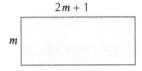

2   Work out an expression for the area of each rectangle in question 1.

3   Work out an expression for the perimeter of each shape.

a 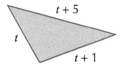   **b**    **c**

d 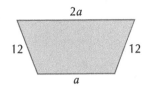   **e**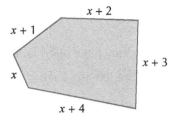

4   Show that the given expressions are equivalent expressions for the perimeters of the shapes in question 3.

a $3(t + 2)$ **b** $4(k + 2)$ **c** $4(c + 4)$ **d** $3(a + 8)$ **e** $5(x + 2)$

5   Write down an expression for the volume of each cuboid.

a 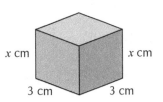   **b**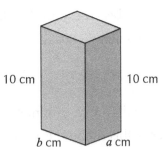

c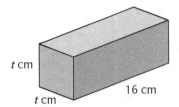

6   In the diagram, all lengths are in centimetres.

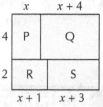

   a   Work out an expression for the area, in cm², of:
       i   rectangle P   ii   rectangle Q
       iii rectangle R   iv   rectangle S.
   b   Show that the total area of rectangles P, Q, R and S is $12(x + 2)$ cm².

   c   Work out an expression for the perimeter, in cm, of:
       i   rectangle P   ii   rectangle Q
       iii rectangle R   iv   rectangle S.
   d   Show that the perimeter of the whole shape is $4(x + 5)$ cm.

7   In the diagram, all lengths are in centimetres.

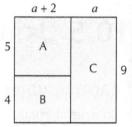

   a   Work out an expression for the area, in cm², of:
       i   rectangle A   ii   rectangle B   iii   rectangle C
       iv the whole shape.
   b   Work out an expression for the perimeter, in cm, of:
       i   rectangle A   ii   rectangle B   iii   rectangle C
       iv the whole shape.

8   This shape has been divided into rectangles in two different ways.

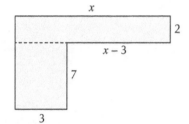 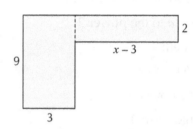

   a   Use the first diagram to write an expression for the area of the whole shape.
   b   Use the second diagram to write a different expression for the area of the whole shape.
   c   Show that the expressions are equivalent.

(PS) 9   This is an algebra wall.

   The expression in each brick is the sum of the expressions in the two bricks below it. For example, the sum of $x + 3$ and $2x + 5$ is $3x + 8$.

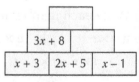

   Show that the expression in the top brick can be written as $6(x + 2)$.

(PS) 10   Show that the expression in the top brick of this algebra wall can be written as $8(x + 2)$.

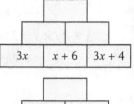

(PS) 11   Show that the expression in the top brick of this algebra wall can be written as $5(x + 3)$.

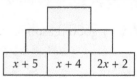

12 Show that the expression in the top brick of this algebra wall can be written as $6(x-1)$.

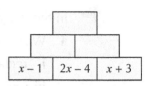

| $x-1$ | $2x-4$ | $x+3$ |

Challenge: Fill in the bricks

Here is an algebra wall.

Find three expressions to put in the bottom row of bricks.

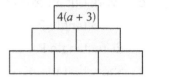

$4(a+3)$

10.5 Using index notation

Learning objective

- To write algebraic expressions involving powers

Key words

| index | power |

You know that you can write $a + a + a$ as $3a$.

In a similar way, you can write $a \times a \times a$ as a^3.

This is called 'a cubed' or 'a to the **power** 3'.

So if the value of a is 4, then:

$$a^3 = 4 \times 4 \times 4 = 64$$

This is 4 cubed or 4 to the power 3.

You can have powers larger than 3.

$$a \times a \times a \times a = a^4$$

This is a to the power 4. The number 4 is called the **index**.

Example 6

Write each number in index notation.

a $t \times t^2$ **b** $n \times 2n \times 4n$ **c** $x \times 2y \times x$

   a $t \times t^2 = t \times t \times t$ Because $t^2 = t \times t$.

        $= t^3$

   b $n \times 2n \times 4n = 2 \times 4 \times n \times n \times n$ You can change the order. Put the numbers together.

           $= 8n^3$ $2 \times 4 = 8$ and $n \times n \times n = n^3$.

   c $x \times 2y \times x = 2 \times x \times x \times y$ Put the number in front.

          $= 2x^2y$ Leave out the $\times$ signs. You could also write it as $2yx^2$.

A very good reason for writing $n \times 3$ as $3n$ and not as $n3$ is so that you will not mistake it for n^3.

Exercise 10E

1 Write each expression in index form.

    a $a \times a \times a \times a$ **b** $r \times r \times r$ **c** $b \times b \times b \times b \times b$

    d $m \times m \times m \times m \times m \times m$ **e** $4a \times 3a$ **f** $p \times 2p$

    g $2g \times 3g \times 2g$ **h** $k \times 4 \times 2k \times k \times 3k$

2 Write each expression as briefly as possible.

    a $f + f + f + f + f$ **b** $w \times w \times w \times w$ **c** $c + c + c + c + c + c + c$

    d $k \times k \times k \times k$ **e** $D + D + D + D + D + D$

3 Explain the difference between $5j$ and j^5.

4 Copy and complete this table.

n	2	4	5
$3n$			
n^3			
$3n^3$			

5 Simplify each of these expressions.

    a $a \times b \times a$ **b** $x \times y \times y \times 6$ **c** $t \times u \times t \times u$ **d** $d \times 2d \times c$

    e $a \times 2b \times b$ **f** $w \times x \times 2w$ **g** $2t \times 2t \times u$ **h** $\frac{1}{2}e \times 3f \times 4e$

6 Given that $f = 5$, work out the value of: **a** $3f^2$ **b** $2f^3$.

7 Write down an expression for the volume of each cuboid.

a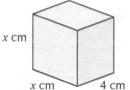

x cm, x cm, 4 cm

b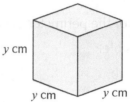

y cm, y cm, y cm

c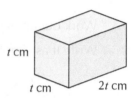

t cm, t cm, $2t$ cm

d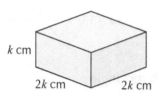

k cm, $2k$ cm, $2k$ cm

e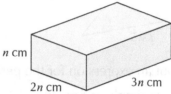

n cm, $2n$ cm, $3n$ cm

8 Write each expression as simply as possible.

    a $t \times t^3$ **b** $2t \times 3t^3$ **c** $t^2 \times t^2$ **d** $5t^2 \times 6t^2$

Challenge: Matching multiples

A Copy and complete this multiplication table.

B Which multiplications in the table give the same answer?

$\times$	a	b	a^2	b^2	ab
2				$2b^2$	
a		ab	a^3		
b					

Ready to progress?

Review questions

1 Write these expressions as simply as possible.

   a $2 + t \times 3$ b $4 \times d \times 5$ c $2 \times a + 4 \times b$ d $c \times c \times \frac{1}{2}$

   e $r \times 2 - 12$ f $20 - 3 \times q$ g $2 \times a \div b$ h $k \times 6 \times m$

2 Write each expression as simply as possible.

   a $h + h + 3h$ b $2a + 4b - a - b$ c $6p + 8 - 2p - 10$ d $3x - 3 + 4x - 5$

   e $x^2 + 5x^2 - 2x^2$ f $4 - a + 3 - a$ g $3d + d^2 - d + 2d^2$ h $2 + r - 10 + 5r^2$

3 Work out an expression for the perimeter of each triangle.

   Write it as simply as possible.

   a   b   c

   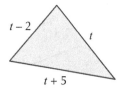   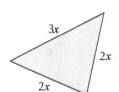   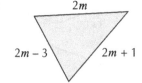

4 Work out an expression for the perimeter of each rectangle.

   Write it as simply as possible.

   a   b   c

   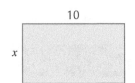   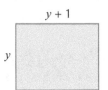   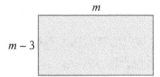

5 Write an expression for the area of each rectangle in question **4**.

6 a Write down an expression for the volume of this cuboid.

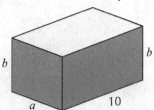

b Two of the faces of the cuboid are red. Work out an expression for the total area of these two faces.

7 Multiply out the brackets in each expression.

a $2(a + 8)$ b $3(f - 6)$ c $\frac{1}{4}(t + 20)$

  8 The mean of three numbers, a, b and c, is $\dfrac{a + b + c}{3}$.

a Write an expression for the mean of four numbers, w, x, y and z.
b Work out the mean if $w = 12$, $x = 10$, $y = 11$ and $z = 47$.

  9 The lengths of the sides of a cuboid are t, $2t$, and $3t$.

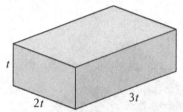

Work out an expression for the volume of the cuboid.

  10 This is the net for a cuboid with a square base.

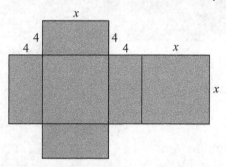

a Work out an expression for the volume of the cuboid.
b Work out an expression for the perimeter of the net.
c Work out an expression for the total area of the net.

Mathematical reasoning
Writing in algebra

- To find the perimeter of a rectangle you add together the four sides of the rectangle or, alternatively, you can just add the length and the width and then multiply the result by 2.

And here is the same information, translated into algebra.

- Perimeter of a rectangle = $2l + 2w$ or $2(l + w)$ where l and w are the width and length of the rectangle.

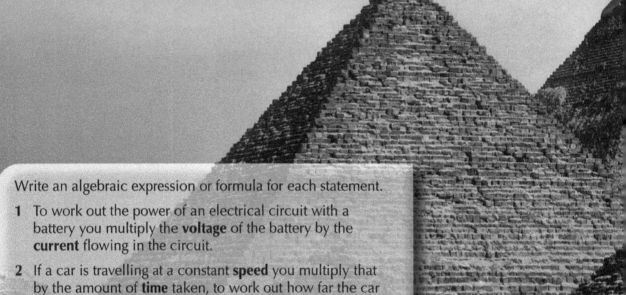

Write an algebraic expression or formula for each statement.

1 To work out the power of an electrical circuit with a battery you multiply the **voltage** of the battery by the **current** flowing in the circuit.

2 If a car is travelling at a constant **speed** you multiply that by the amount of **time** taken, to work out how far the car has travelled.

3 To find the volume of a square-based pyramid (like those built by the Egyptians) you multiply the length of the **side** of the square base by itself, multiply that by the **height** of the pyramid, and then multiply that by $\frac{1}{3}$.

4 To find the area of a regular pentagon, multiply the length of a **side** by itself and then multiply the answer by 1.72.

5 To work out the volume of a cylindrical can, multiply the **diameter** of the top by itself, multiply the answer by the **height** of the can, and then multiply by 0.79.

6 You can find the height of a building by dropping a coin from the top. The height, in metres, is the **time**, in seconds, it takes to reach the ground, multiplied by itself and then multiplied by 5.

7 To work out a person's body mass index you square the person's **height**, in metres, and then divide the person's **mass** in kilograms by that number.

8 To find the kinetic energy of a moving object you multiply the **speed** of the object by itself, then multiply by the **mass** of the object, then halve the answer.

9 To work out the cost of your electricity bill for a 90-day period, you first multiply the **daily charge** by 90. Then you multiply the **number of kilowatt-hours** of electricity you have used by the **cost per kilowatt-hour** and add that to the previous result.

11

Congruence and scaling

This chapter is going to show you:

- how to recognise congruent shapes
- how to enlarge a shape by a scale factor
- how to use shape and ratio
- how to use scales in drawings and maps.

You should already know:

- how to reflect a 2D shape in a mirror line
- how to rotate a 2D shape about a point
- how to translate a 2D shape
- how to use ratio.

About this chapter

In a golden rectangle, the side lengths are in the ratio $1 : \phi$. ϕ is the Greek letter phi and is equal to $\frac{\sqrt{5}+1}{2} = 1.618\,033\,988$ $749\,894\,848\,204\,586\,834\,365\,638\,117\,720\,309\,179\,805\,76\ldots$ or 1.618 for short!

This rectangle is special, because if you cut a square from one end of it, you will be left with a smaller shape that is another golden rectangle, with sides that are in the same ratio as the one you started with.

The golden rectangle has been described as one of the most visually pleasing rectangular shapes. Many artists and architects have used golden rectangles in their work.

For example, Leonardo da Vinci's *Mona Lisa* and Salvador Dali's *The Sacrament of the Last Supper* are golden rectangles and the exterior of the Parthenon on the Acropolis in Athens is based on the golden rectangle.

The flag of Togo was designed to approximate a golden rectangle. The same shape can also be seen in flower seed heads, and in the shells of sea creatures.

11.1 Congruent shapes

Learning objective

* To recognise congruent shapes

Key words

congruent	congruent triangles
transformation	

All the triangles on this grid are reflections, rotations or translations of triangle A. What do you notice about them?

You should remember that, for reflections, rotations and translations, the image is always exactly the same shape and size as the object.

If two shapes are exactly the same shape and size, they are **congruent**. Reflections, rotations and translations all produce images that are congruent to the original object. For shapes that are congruent, all the corresponding sides are equal and all the corresponding angles are equal.

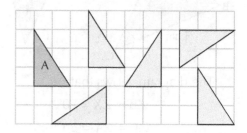

Example 1

Which two shapes below are congruent?

a   b   c   d

Shapes **b** and **d** are exactly the same shape and size, so they are congruent.

Use tracing paper to check that the two shapes are congruent.

Congruent triangles

Two triangles are congruent if:

* all three sides are the same lengths in both triangles (SSS)

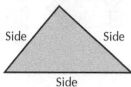

* two sides are the same length and the angle between them is the same size in both triangles (SAS)

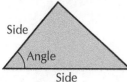

* two angles are the same size and the side between them is the same length in both triangles (ASA).

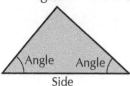

Example 2

Show that triangle ABC is congruent to triangle XYZ.

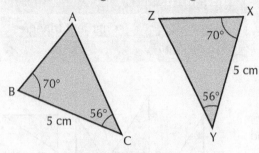

The diagram shows that:

$$\angle B = \angle X = 70°$$
$$\angle C = \angle Y = 56°$$
$$BC = XY = 5 \text{ cm}$$

So, triangle ABC is congruent to triangle XYZ (ASA).

Exercise 11A

1. Look at the shapes in each pair and state whether they are congruent or not. Use tracing paper to help if you are not sure.

a    b    c

d 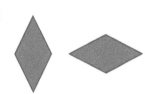   e 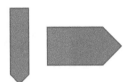   f

2. Which pairs of shapes on the grid below are congruent?

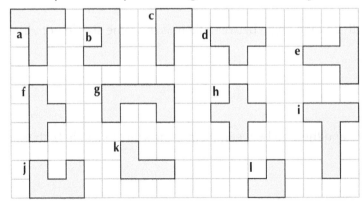

3 Which of the shapes below are congruent?

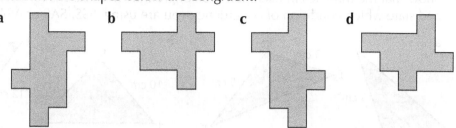

a   b   c   d

4 Two congruent right-angled triangles are placed together with two of their equal sides touching to make another shape, as shown in the diagram.

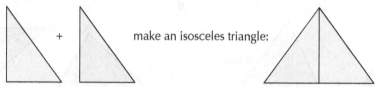

+   make an isosceles triangle:

a Using the same triangles, how many different shapes can you make? To help you find out, cut out the triangles from a piece of card.

b Repeat the activity using two congruent isosceles triangles.

c Repeat the activity using two congruent equilateral triangles.

(MR) **5** This is a right-angled triangle.

Show how it can be split up into four congruent right-angled triangles.

(MR) **6** Show how this shape can be split into three congruent shapes.

(PS) **7** This 4 by 4 pinboard is divided into two congruent shapes.

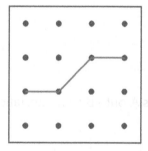

a Use square-dotted paper to show different ways this can be done.

b Can you divide the pinboard into four congruent shapes?

8 Show that the triangles in each pair are congruent. Give reasons for each answer and state which condition of congruency you are using: SSS, SAS or ASA.

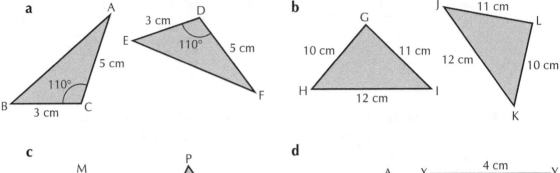

a

b

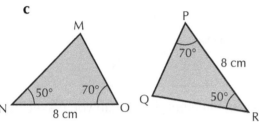
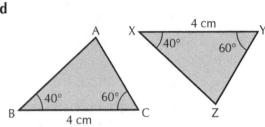

c

d

Reasoning: Combined transformations

A **transformation** is a way of changing the position of a shape.

Examples of transformations are reflections, rotations and translations.

A combined transformation is a way of changing a shape by using two or more different transformations.

Copy the congruent T-shapes A, B, C and D onto a square grid, as shown.

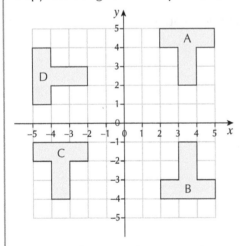

For example, a combination of two transformations that moves A onto B is: a translation of 1 unit down followed by a reflection in the *x*-axis.

Find a combination of two transformations that will move:

a A onto C   **b** A onto D   **c** B onto C

d B onto D   **e** C onto D.

11.2 Enlargements

Learning objective

* To enlarge a 2D shape by a scale factor

Key words

| centre of enlargement |
| enlargement |
| ray |
| scale factor |

The three transformations you have met so far, reflections, rotations and translations, do not change the size of the object. You are now going to look at a transformation that does change the size of an object: an **enlargement**. The illustration shows a photograph that has been enlarged.

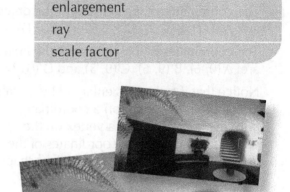

The diagram shows triangle ABC enlarged to give triangle A'B'C'.

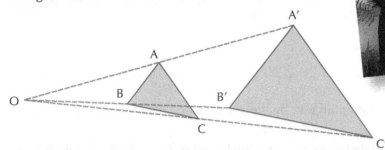

All the sides of triangle A'B'C' are twice as long as the sides of triangle ABC. O is called the **centre of enlargement**. The dotted lines from O are called the guidelines or **rays** for the enlargement.

Notice that OA' = 2 × OA, OB' = 2 × OB and OC' = 2 × OC.

Triangle ABC is enlarged by a **scale factor** of two about the centre of enlargement O to give the image triangle A'B'C'.

To enlarge a shape, you need a centre of enlargement and a scale factor.

Example 3

Enlarge the triangle XYZ by a scale factor of two about the centre of enlargement O.

Draw **rays** OX, OY and OZ. Measure the length of the three rays and multiply each of these lengths by two.

Then extend each of the rays to these new lengths, measured from O, and plot the points X', Y' and Z'.

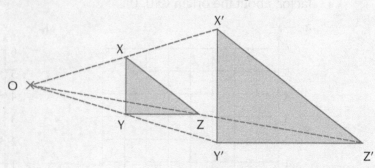

Join X', Y' and Z'.

Triangle X'Y'Z' is the enlargement of triangle XYZ by a scale factor of two about the centre of enlargement O.

Example 4

The rectangle ABCD on this coordinate grid has been enlarged by a scale factor of three about the origin O(0, 0) to give the image rectangle A'B'C'D'.

Compare the coordinates of the vertices of the object and the image. What do you notice?

The coordinates of the vertices of the object are: A(0, 2), B(3, 2), C(3, 1) and D(0, 1).

The coordinates of the vertices of the image are: A'(0, 6), B'(9, 6), C'(9, 3) and D'(0, 3).

Notice that if a shape is enlarged by a scale factor about the origin on a coordinate grid, the coordinates of a vertex on the enlarged shape are the coordinates of the corresponding vertex on the original shape, multiplied by the scale factor.

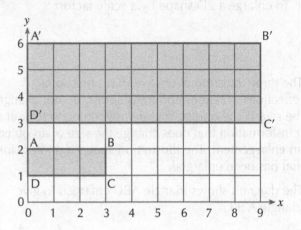

Exercise 11B

1   Copy or trace each shape and enlarge it, by the given scale factor, about the centre of enlargement O.

   a   Scale factor 2   b   Scale factor 3   c   Scale factor 2   d   Scale factor 3

(**Note:** × is the centre of square)

2   Copy each diagram onto centimetre-squared paper and enlarge it, by the given scale factor, about the origin O(0, 0).

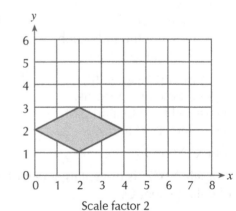

a   Scale factor 2

b   Scale factor 2

c

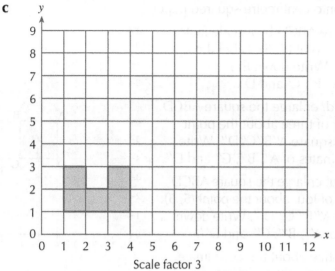

Scale factor 3

d

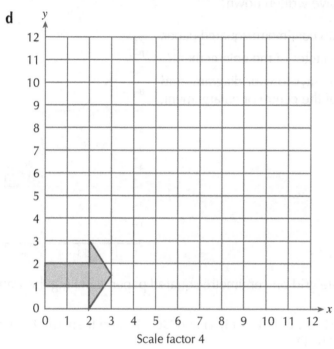

Scale factor 4

3 Draw a coordinate grid on centimetre-squared paper, labelling the x and y axes from 0 to 12.

Plot the points A(4, 6), B(5, 4), C(4, 1) and D(3, 4) and join them together to form the kite ABCD. Enlarge the kite by a scale factor of 2 about the origin O(0, 0).

(MR) (4) Copy the diagram onto centimetre-squared paper.

a Enlarge the square ABCD by a scale factor of two about the point (5, 5). Label the square A′B′C′D′. Write down the coordinates of A′, B′, C′ and D′.

b On the same grid, enlarge the square ABCD by a scale factor of three about the point (5, 5). Label the square A″B″C″D″. Write down the coordinates of A″, B″, C″ and D″.

c On the same grid, enlarge the square ABCD by a scale factor of four about the point (5, 5). Label the square A‴B‴C‴D‴. Write down the coordinates of A‴, B‴, C‴ and D‴.

d What do you notice about the coordinate points that you have written down?

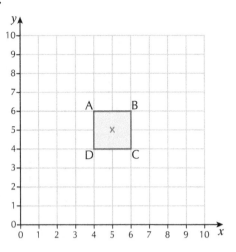

(5) Copy the diagram onto centimetre-squared paper.

a What is the scale factor of the enlargement?

b By adding suitable rays to your diagram, find the coordinates of the centre of enlargement.

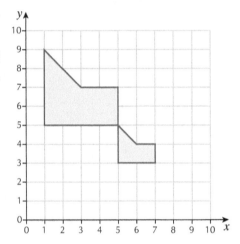

(MR) (6) a Draw a coordinate grid on centimetre-squared paper, labelling the x and y axes from 0 to 12.

Plot the points A(1, 3), B(3, 3), C(3, 1) and D(1, 1), and then join them together to form the square ABCD.

b Write down the area of the square.

c Enlarge the square ABCD by a scale factor of 2 about the origin. What is the area of the enlarged square?

d Enlarge the square ABCD by a scale factor of 3 about the origin. What is the area of the enlarged square?

e Enlarge the square ABCD by a scale factor of 4 about the origin. What is the area of the enlarged square?

f Write down anything you notice about the increase in area of the enlarged squares. Try to write down a rule to explain what is happening.

g Repeat the above using your own shapes. Does your rule still work?

Activity: Enlarged stickmen

Work in a pair or a group.

Design a poster to show how the 'stickman' shown can be enlarged by different scale factors about any convenient centre of enlargement.

11.3 Shape and ratio

Learning objective

- To use ratio to compare lengths, areas and volumes of 2D and 3D shapes

Key word

hectare

You can use ratio to compare lengths, areas and volumes of 2D and 3D shapes, as these examples show.

Example 5

Work out the ratio of the length of the line AB to the length of the line CD.

A ———— B   C ——————————————— D
12 mm   4.8 cm

First convert the measurements to the same units and then simplify the ratio.

Always use the smaller unit, which here is millimetres (mm).

    4.8 cm = 48 mm

Then the ratio is 12 mm : 48 mm = 1 : 4.

 Hint Remember that ratios do not have units.

Example 6

Work out the ratio of the area of rectangle A to the area of rectangle B, giving the answer in its simplest form.

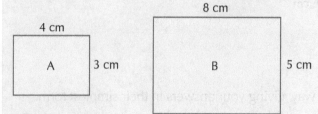

The ratio is:

    12 cm² : 40 cm²

    = 3 : 10

Example 7

Work out the ratio of the volume of the
cube to the volume of the cuboid, giving
the answer in its simplest form.

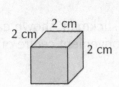

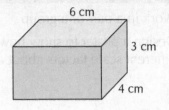

The ratio is:

8 cm³ : 72 cm³

= 1 : 9

Exercise 11C

1   Express each ratio in its simplest form.

   a   10 mm : 25 mm   **b**   2 mm : 2 cm   **c**   36 cm : 45 cm

   d   40 cm : 2 m   **e**   500 m : 2 km

2   Look at the two squares then work out each ratio, giving your answers in their
   simplest form.

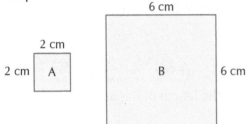

   a   The length of a side of square A to the length of a side of square B

   b   The perimeter of square A to the perimeter of square B

   c   The area of square A to the area of square B

3   Three rectangles A, B and C are arranged as in the diagram. The ratio of the length
   of rectangle A to the length of rectangle B to the length of rectangle C is
   3 cm : 6 cm : 9 cm = 1 : 2 : 3.

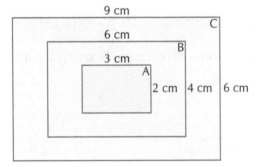

   a   Work out each ratio in the same way, giving your answers in their simplest form.

   i   The width of rectangle A to the width of rectangle B to the width of rectangle C

   ii The perimeter of rectangle A to the perimeter of rectangle B to the perimeter of
           rectangle C

   iii The area of rectangle A to the area of rectangle B to the area of rectangle C

   b   Write down anything you notice about the three rectangles.

4 In the diagram, Flag X is mapped onto Flag Y by a reflection in mirror line 1. Flag X is also mapped onto Flag Z by a reflection in mirror line 2.

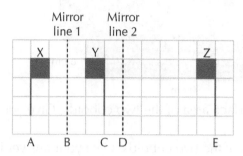

Work out the ratios of these lengths, giving your answers in their simplest form.

a AB : BC   **b** AB : AE   **c** AC : AE   **d** BD : CE

5 **a** Work out the ratio of the area of the orange square to the area of the yellow surround, giving your answer in its simplest form.

     b Express the area of the orange square as a fraction of the area of the yellow surround.

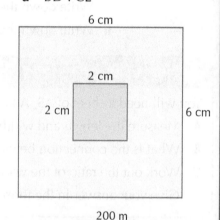

6 The dimensions of lawn A and lawn B are given on the diagram.

     a Calculate the area of lawn A, giving your answer in square metres.

     b Calculate the area of lawn B, giving your answer in:

        i square metres   **ii** hectares. (1 hectare = 10 000 m²)

     c Work out the ratio of the length of lawn A to the length of lawn B, giving your answer in its simplest form.

     d Work out the ratio of the area of lawn A to the area of lawn B, giving your answer in its simplest form.

     e Express the area of lawn A as a fraction of the area of lawn B.

7 The dimensions of a fish tank are given on this diagram.

     a Calculate the volume of the fish tank, giving your answer in litres. (1 litre = 1000 cm³)

     b The fish tank is filled with water to a depth of $\frac{3}{4}$ of the height. Calculate the volume of water in the fish tank, giving your answer in litres.

     c Work out the ratio of the volume of water in the fish tank to the total volume of the fish tank, giving your answer in its simplest form.

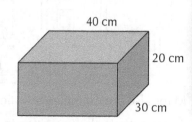

8 **a** Work out the ratio of the sides of the two cuboids. Give your answer in the form $1 : n$.

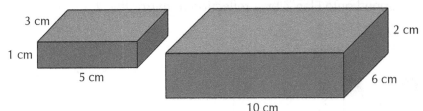

b Work out the ratio of the total surface area of the two cuboids. Give your answer in the form $1 : n$.

c Work out the ratio of the volume of the two cuboids. Give your answer in the form $1 : n$.

d The ratio of the sides of two cuboids is $1 : 3$.

   i Write down the ratio of the total surface area of the two cuboids.

   ii Write down the ratio of the volume of the two cuboids.

Activity: Paper sizes

You will need a sheet of A5, A4 and A3 paper for this activity.

A Measure the length and width of the sides of each sheet of paper to the nearest millimetre.

B What is the connection between the length and width of successive paper sizes?

C Work out the ratio of the width to the length for each successive paper size.

Give your answer in the form $1 : n$. What do you notice?

11.4 Scales

Learning objectives

- To understand and use scale drawings
- To know how to use map ratios

Key words	
map ratio	scale
scale drawing	

A **scale drawing** is a smaller drawing of an actual object. A **scale** must always be clearly given by the side of or below the scale drawing.

This is part of an architect's blue-print. It is a scale drawing for an extension to a building.

Maps also are examples of scale drawings. On most maps the scale is given as a **map ratio**.

A map ratio is always written in terms of the same units. So, for example, a scale of 1 cm : 1 km would be 1 : 100 000 in a map ratio.

Example 8

The diagram is a scale drawing of Rebecca's room.

What are the real measurements of the room?

- On the scale drawing, the length of the room is 5 cm, so the actual length of the room is 5 m.
- On the scale drawing, the width of the room is 3.5 cm, so the actual width of the room is 3.5 m.
- On the scale drawing, the width of the window is 2 cm, so the actual width of the window is 2 m.

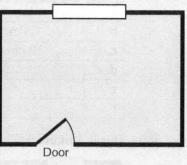

Door

Scale: 1 cm to 1 m

Example 9

The scale of a map is 1 cm to 5 km.

Work out the scale as a map ratio.

1 m = 100 cm and 1 km = 1000 m

5 km = 5 × 1000 × 100 = 500 000 cm

So 5 km = 500 000 cm.

The map ratio is 1 : 500 000.

Exercise 11D

1. The lines shown are drawn using a scale of 1 cm to 10 m.
    Write down the length each line represents.

    a ────────
    b ──────────────────
    c ──────────────
    d ────────────────────────
    e ──────────────────

2. The diagram shows a scale drawing for a school hall.
    a Find the actual length of the hall.
    b Find the actual width of the hall.
    c Find the actual distance between the opposite corners of the hall.

    Scale: 1 cm to 5 m

    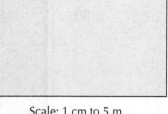

3. The diagram shown is Ryan's scale drawing for his mathematics classroom. Nathan notices that Ryan has not put a scale on the drawing, but he knows that the length of the classroom is 8 m.

    a What scale has Ryan used?
    b What is the actual width of the classroom?
    c What is the actual area of the classroom?

4 Copy and complete this table for a scale drawing in which the scale is 4 cm to 1 m.

	Actual length	Length on scale drawing
a	4 m	
b	1.5 m	
c	50 cm	
d		12 cm
e		10 cm
f		4.8 cm

5 This is a plan for a bungalow.

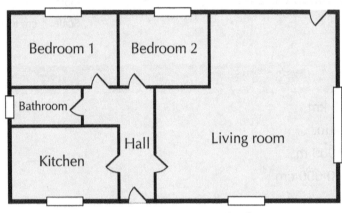

Scale: 1 cm to 2 m

a Find the actual dimensions of:

   i the kitchen   **ii** the bathroom   **iii** bedroom 1   **iv** bedroom 2.

b Calculate the actual area of the living room.

 6 The diagram shows the plan of a football pitch. It is not drawn to scale. Use the measurements on the diagram to make a scale drawing of the pitch. Choose your own scale.

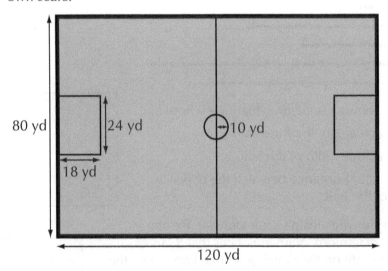

7 The map shows York city centre.

Scale: 1 : 25 000

Find the actual direct distance between:

a the Law courts and the station

b the Castle Museum and the station

c the Castle Museum and the football ground

d the National Railway Museum and the Castle Museum.

Reasoning: Map ratios

Write each scale as a map ratio.

A 1 cm to 100 m   **B** 4 cm to 500 m   **C** 1 cm to 1 km

D 2 cm to 1 km   **E** 1 cm to 10 km

Ready to progress?

I can recognise congruent shapes.
I know how to make a scale drawing.

I can enlarge a 2D shape by a scale factor about a centre of enlargement.
I can use ratio to compare lengths, areas and volumes of 2D and 3D shapes.
I can use map ratios.

I can recognise congruent triangles.

Review questions

1 a Four congruent trapezia can join to make a parallelogram. Copy the diagram and draw two more trapezia to complete the drawing of the bigger parallelogram.

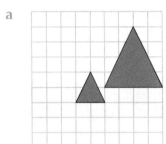

   b Four congruent trapezia can join together to make a bigger trapezium. Copy the diagram and draw two more trapezia to complete the drawing of the bigger trapezium.

2 Copy each diagram. Mark the centre of enlargement with a cross.

   a

   b

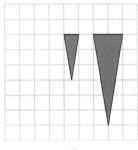

3 Copy the grid. On your copy, draw an enlargement of scale factor two of the trapezium. Use point C as the centre of enlargement.

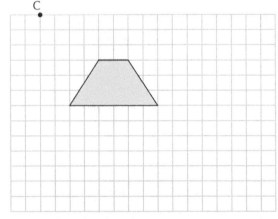

4 a Work out the ratio of the area of the red squares to the area of the whole square, giving your answer in its simplest form.

b What percentage of the shape is shaded blue?

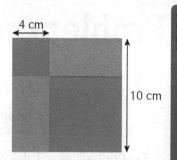

4 cm

10 cm

5 This is a map of Wales.

Work out the actual direct distance between:

a Cardiff and Bangor

b Holyhead and Newport

c Milford Haven and Abergavenny

d Aberystwyth and Milford Haven.

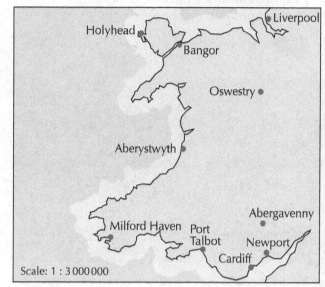

Scale: 1 : 3 000 000

6 The surface area of this cuboid is 118 cm².

a Set up an equation in terms of x.

b Solve the equation.

c Work out the volume of the cuboid.

d The volume of the cuboid is increased by 15%. Work out the volume of the enlarged cuboid.

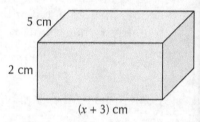

5 cm

2 cm

$(x + 3)$ cm

7 The diagram shows five triangles. All lengths are in centimetres.

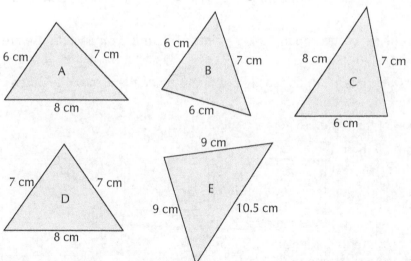

Write down the letters of two triangles that are congruent to each other.

Explain how you know they are congruent.

Problem solving

Photographs

Print size	Price each	
3" × 2" (4)	£0.99	
	Quantity	**Price**
13 cm × 9 cm	1–99	£0.10 each
	100–249	£0.09 each
	250+	£0.08 each
	Quantity	**Price**
6" × 4"	1–49	£0.15 each
	50–99	£0.12 each
	100–249	£0.09 each
	250–499	£0.08 each
	500–750	£0.06 each
	751+	£0.05 each
7" × 5"	£0.29	
8" × 6"	£0.45	
10" × 8"	£1.20	
12" × 8"	£1.20	
45 cm × 30 cm	£6.99	

FastPrint advertises the cost of printing photographs in their shop.

 Hint The " symbol means inches, and 1 inch ≈ 2.5 cm.

1   A school decides to use FastPrint to buy prints of a year group photograph. Pupils can choose the size of the prints they want. They can also choose to buy more than one size. This is the school's order.

128	13 cm × 9 cm prints	75	10" × 8" prints
87	6" × 4" prints	60	12" × 8" prints

What is the total cost of buying these prints?

2   The print sizes are given in both imperial units and metric units.

   a   Use the conversion factor 1 cm = 0.394 inches to change the 13 cm × 9 cm and the 45 cm × 30 cm print sizes into imperial sizes. Give your answers correct to one decimal place.

   b   Find the area of each of the imperial sized prints. (The units you need to use are square inches or sq in.)

   c   There are pairs of prints in which the area of one is twice the area of the other. Which are they?

3   These are the sizes of three picture frames: A, B and C.

   a   The 7" × 5" print will fit inside frame A. What will be the area of the outside border?

   b   Which of the prints will best fit inside the other two frames, if a suitable border is to be left around the print?

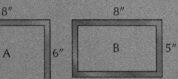

4 Some of the prints are actual mathematical enlargements of each other.

Write down the sizes of the prints that are exact enlargements of each other and state the scale factor of the enlargement.

12″ × 8″  10″ × 8″

8″ × 6″  7″ × 5″   6″ × 4″

5 EasyPrint also advertises the cost of photograph prints in their shop.

3″ × 2″ print	£0.25 each	8″ × 6″ print	£0.42 each
6″ × 4″ print	£0.12 each	10″ × 8″ print	£1.20 each
7″ × 5″ print	£0.20 each	12″ × 8″ print	£1.32 each

a If you order one of each print size from EasyPrint, which prints are cheaper than FastPrint?

b If you wanted to order 120 6″ × 4″ prints, which shop would you choose? How much would you save?

c How much more do you pay for a 12″ × 8″ print at EasyPrint?

d What is the percentage increase in the price if you ordered 12″ × 8″ prints from EasyPrint rather than from FastPrint?

6 Any rectangle with length and width in the ratio 1.618 : 1 is known as a golden rectangle.

a Work out the ratio of the length to the width, in the form n : 1, for each print size at EasyPrint.

For example, for the 3″ × 2″ print, 3 : 2 = 1.5 : 1.

b Which of the prints are close to being golden rectangles?

12

Fractions and decimals

This chapter is going to show you:

- how to multiply fractions and integers
- how to divide fractions and integers
- how multiply large and small numbers
- how to divide large and small numbers.

You should already know:

- the relationship between mixed numbers and improper fractions
- how to add two fractions or mixed numbers
- how to subtract two fractions or mixed numbers
- how to find the lowest common multiple (LCM).

About this chapter

Fractions have been written in different ways during human history. Nowadays we use two different ways of writing fractional numbers – either as one whole number over another whole number, or with a decimal point.

Why do you need to know both? Decimals are often more convenient but there are some simple fractions that are not easy to write as decimals. For example:

- $\frac{2}{3}$ as a decimal is 0.666 666 666...
- $\frac{1}{7}$ as a decimal is 0.142 857 142...

Because of this it is useful to be able to write and calculate with fractional numbers in both ways.

In this chapter you will be learning to do calculations with fractions and with decimals. In particular, you will be looking at multiplication and division.

12.1 Adding and subtracting fractions

Learning objective

• To add and subtract fractions and mixed numbers

You should already know how to add and subtract fractions.

To do this, you need to make sure that all of the fractions have the same denominator. This means that you have to change one or more of the fractions to equivalent fractions.

Start by finding the lowest common multiple of the two denominators of the fractions you start with. This can be the new denominator for both fractions.

Read through these examples, to remind you what to do.

Example 1

Add these numbers. **a** $\frac{3}{4} + \frac{5}{6}$ **b** $2\frac{7}{8} + 3\frac{1}{4}$

a $\frac{3}{4} + \frac{5}{6}$ Change the denominators to 12, because this is the lowest

                         common multiple of 4 and 6.

$\frac{3}{4} + \frac{5}{6} = \frac{9}{12} + \frac{10}{12}$ $\frac{3}{4} = \frac{3 \times 3}{4 \times 3} = \frac{9}{12}$ and $\frac{5 \times 2}{6 \times 2} = \frac{10}{12}$

$= \frac{19}{12}$

$= 1\frac{7}{12}$ $19 \div 12 = 1$ remainder 7

b $2\frac{7}{8} + 3\frac{1}{4} = 2 + 3 + \frac{7}{8} + \frac{1}{4}$

        $= 5 + \frac{7}{8} + \frac{2}{8}$ $\frac{1 \times 2}{4 \times 2} = \frac{2}{8}$

        $= 5 + \frac{9}{8}$

        $= 5 + 1\frac{1}{8}$ $\frac{9}{8} = 1\frac{1}{8}$

        $= 6\frac{1}{8}$

Example 2

Calculate: **a** $\frac{5}{8} - \frac{1}{3}$ **b** $3\frac{1}{2} - 1\frac{7}{8}$ **c** $13\frac{3}{4} - 11\frac{1}{6}$.

a $\frac{5}{8} - \frac{1}{3} = \frac{15}{24} - \frac{8}{24}$ The lowest common multiple of 8 and 3 is 24.

      $= \frac{7}{24}$ This cannot be simplified.

b $3\frac{1}{2} - 1\frac{7}{8} = \frac{7}{2} - \frac{15}{8}$ Change to improper fractions.

       $= \frac{28}{8} - \frac{15}{8}$ 2 is a factor of 8 so change $\frac{7}{2}$ to eighths.

       $= \frac{13}{8}$ $28 - 15 = 13$

       $= 1\frac{5}{8}$ $13 \div 8 = 1$ remainder 5

(continued)

c $13\frac{3}{4} - 11\frac{1}{6} = 13 - 11 + \frac{3}{4} - \frac{1}{6}$ You can separate the integer and fraction parts.

$= 2 + \frac{9}{12} - \frac{2}{12}$ 12 is the lowest common multiple of 4 and 6 so change to twelfths.

$= 2\frac{7}{12}$ $\frac{9}{12} - \frac{2}{12} = \frac{7}{12}$

Exercise 12A

1 Add these fractions. Give each answer as simply as possible, as a mixed number if necessary.

a $\frac{1}{4} + \frac{5}{8}$ **b** $\frac{3}{4} + \frac{5}{12}$ **c** $\frac{5}{6} + \frac{2}{3}$ **d** $\frac{3}{5} + \frac{9}{10}$

e $\frac{2}{3} + \frac{3}{4}$ **f** $\frac{3}{5} + \frac{1}{2}$ **g** $\frac{7}{8} + \frac{1}{3}$ **h** $\frac{5}{6} + \frac{1}{4}$

2 Add these fractions. Give each answer as simply as possible, as a mixed number if necessary.

a $1\frac{1}{4} + \frac{3}{8}$ **b** $\frac{3}{4} + 1\frac{7}{8}$ **c** $1\frac{1}{6} + 1\frac{2}{3}$ **d** $2\frac{3}{10} + 1\frac{1}{4}$

e $1\frac{1}{3} + 2\frac{3}{4}$ **f** $1\frac{5}{8} + 4\frac{1}{12}$ **g** $8\frac{1}{2} + 5\frac{1}{3}$ **h** $2\frac{5}{6} + 2\frac{3}{8}$

3 Subtract these fractions. Write the answers as simply as possible.

a $\frac{2}{3} - \frac{1}{6}$ **b** $\frac{2}{3} - \frac{5}{12}$ **c** $\frac{3}{4} - \frac{1}{8}$ **d** $\frac{7}{10} - \frac{1}{4}$

e $\frac{7}{8} - \frac{5}{12}$ **f** $\frac{2}{3} - \frac{1}{5}$ **g** $\frac{3}{4} - \frac{1}{10}$ **h** $\frac{9}{10} - \frac{2}{3}$

4 Subtract these fractions. Write the answers as simply as possible.

a $2 - \frac{3}{8}$ **b** $1\frac{2}{3} - \frac{1}{6}$ **c** $2\frac{1}{4} - \frac{5}{8}$ **d** $1\frac{3}{8} - \frac{3}{4}$

e $3\frac{1}{2} - 1\frac{3}{4}$ **f** $3\frac{2}{3} - 1\frac{1}{2}$ **g** $3\frac{3}{4} - 1\frac{5}{12}$ **h** $4\frac{1}{10} - 1\frac{2}{5}$

5 Subtract these mixed numbers.

a $10\frac{3}{4} - 10\frac{1}{2}$ **b** $9\frac{2}{3} - 8\frac{1}{3}$ **c** $7\frac{1}{4} - 6\frac{3}{8}$ **d** $12\frac{1}{8} - 10\frac{3}{4}$

(PS) 6 Find the missing number in each addition

a $1\frac{1}{4} + \ldots = 2\frac{3}{4}$ **b** $\frac{3}{4} + \ldots = 2\frac{1}{2}$ **c** $1\frac{1}{5} + \ldots = 4\frac{1}{2}$ **d** $2\frac{7}{8} + \ldots = 5\frac{1}{4}$

(PS) 7 Find the missing number in each subtraction.

a $\ldots - \frac{5}{8} = 2$ **b** $\ldots - 1\frac{1}{2} = 1\frac{1}{4}$ **c** $\ldots - \frac{5}{8} = 2\frac{1}{2}$ **d** $\ldots - 4\frac{1}{2} = 1\frac{3}{10}$

(PS) 8 Calculate the perimeter of each rectangle.

a

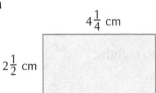

$4\frac{1}{4}$ cm

$2\frac{1}{2}$ cm

b

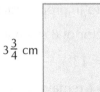

$1\frac{3}{8}$ cm

$3\frac{3}{4}$ cm

c

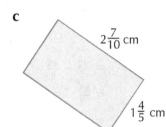

$2\frac{7}{10}$ cm

$1\frac{4}{5}$ cm

9 A knife has a total length of $13\frac{2}{3}$ cm.
The handle is $6\frac{1}{2}$ cm long.
How long is the blade?

$13\frac{2}{3}$ cm

$6\frac{1}{2}$ cm

  10 **a** Work out each answer.

i $\frac{1}{2} - \frac{1}{4}$ **ii** $\frac{1}{2} - \frac{1}{4} + \frac{1}{8}$ **iii** $\frac{1}{2} - \frac{1}{4} + \frac{1}{8} - \frac{1}{16}$

b **i** Predict the answer to: $\frac{1}{2} - \frac{1}{4} + \frac{1}{8} - \frac{1}{16} + \frac{1}{32}$.

ii Check whether your prediction is correct.

  11 **a** Which is larger, $\frac{1}{3}$ or $\frac{3}{8}$? **b** Work out the difference between $\frac{1}{3}$ and $\frac{3}{8}$.

Challenge: Magic square

This is a fractional magic square.

$\frac{2}{15}$		
$\frac{7}{15}$	$\frac{1}{3}$	$\frac{1}{5}$

Every row, column and diagonal has the same total, called the magic number.

A What is the magic number?

B Copy the magic square and fill in the missing fractions.

12.2 Multiplying fractions and integers

Learning objective

• To multiply a fraction and an integer

Look at these three calculations.

$\frac{1}{4}$ of 6 $\qquad$ $\frac{1}{4} \times 6$ $\qquad$ $6 \times \frac{1}{4}$

They all have the same answer.

• $\frac{1}{4}$ of $6 = 6 \div 4 = 1\frac{2}{4} = 1\frac{1}{2}$

• $\frac{1}{4} \times 6 = \frac{1}{4} + \frac{1}{4} + \frac{1}{4} + \frac{1}{4} + \frac{1}{4} + \frac{1}{4} = \frac{6}{4} = 1\frac{2}{4} = 1\frac{1}{2}$

• $6 \times \frac{1}{4}$ is the same as $\frac{1}{4} \times 6$ because the order in which you multiply two numbers does not matter.

Example 3

Work these out. **a** $\frac{2}{3} \times 4$ **b** $7 \times \frac{3}{5}$

a $\frac{2}{3} \times 4 = \frac{2 \times 4}{3}$ Multiply the numerator by 4.

$\qquad = \frac{8}{3} = 2\frac{2}{3}$ $8 \div 3 = 2$ remainder 2. The denominator does not change.

b $7 \times \frac{3}{5} = \frac{21}{5} = 4\frac{1}{5}$ $7 \times 3 = 21$ and $21 \div 5 = 4$ remainder 1.

The next example shows how to multiply a mixed number by an integer.

Example 4

Work out $2\frac{3}{8} \times 4$.

$2\frac{3}{8} \times 4 = (2 \times 4) + (\frac{3}{8} \times 4)$ Think of $2\frac{3}{8}$ as $2 + \frac{3}{8}$ and multiply each term by 4.

$\qquad = 8 + \frac{12}{8}$ $\frac{3}{8} \times 4 = \frac{3 \times 4}{8} = \frac{12}{8}$

$\qquad = 8 + 1\frac{4}{8}$

$\qquad = 9\frac{1}{2}$ $\frac{4}{8} = \frac{1}{2}$

This is similar to what you do in algebra, when you multiply out brackets.

$\qquad 4(2 + f) = 8 + 4f.$

In Example 4, $f = \frac{3}{8}$.

Exercise 12B

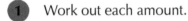

1   Work out each amount.

   a $\frac{1}{3}$ of 7   **b** $\frac{1}{5}$ of 18   **c** $\frac{1}{8}$ of 14   **d** $\frac{1}{6}$ of 16

2   Work these out.

   a $\frac{1}{3} \times 9$ **b** $\frac{1}{5} \times 20$ **c** $\frac{1}{4} \times 10$ **d** $\frac{1}{8} \times 3$

   e $7 \times \frac{1}{4}$ **f** $9 \times \frac{1}{5}$ **g** $15 \times \frac{1}{12}$ **h** $7 \times \frac{1}{14}$

3   Work these out.

   a $\frac{2}{3} \times 2$ **b** $\frac{3}{5} \times 4$ **c** $\frac{3}{4} \times 3$ **d** $\frac{7}{8} \times 2$

   e $4 \times \frac{3}{4}$ **f** $8 \times \frac{2}{5}$ **g** $3 \times \frac{11}{12}$ **h** $6 \times \frac{3}{8}$

4   Work these out.

   a $1\frac{1}{4} \times 3$ **b** $1\frac{1}{4} \times 5$ **c** $1\frac{3}{5} \times 2$ **d** $2\frac{1}{8} \times 4$

   e $4 \times 2\frac{3}{4}$ **f** $7 \times 1\frac{1}{5}$ **g** $4 \times 1\frac{5}{8}$ **h** $3 \times 1\frac{5}{12}$

PS **5** Work out the area of each rectangle.

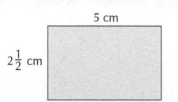

5 cm

$2\frac{1}{2}$ cm

5 cm

$6\frac{1}{4}$ cm

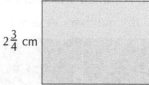

3 cm

$2\frac{3}{4}$ cm

PS **6** This shape is divided into two rectangles.

$5\frac{1}{4}$ cm

3 cm

$2\frac{1}{2}$ cm

2 cm

x cm

a Work out the value of x.

b Work out the area of the shape.

7 Work these out.

a $5\frac{1}{2} \times 7$ **b** $8\frac{1}{4} \times 6$ **c** $2\frac{3}{5} \times 12$ **d** $4\frac{5}{8} \times 20$

e $10\frac{1}{4} \times 10$ **f** $12\frac{2}{3} \times 8$ **g** $3\frac{1}{3} \times 50$ **h** $7\frac{5}{16} \times 11$

8 $66\frac{2}{3}\% = \frac{2}{3}$

Use this fact to work out:

a $66\frac{2}{3}\%$ of 12 **b** $66\frac{2}{3}\%$ of 17 **c** $66\frac{2}{3}\%$ of 31.

9 $37\frac{1}{2}\% = \frac{3}{8}$

Use this fact to work out:

a $37\frac{1}{2}\%$ of 3 **b** $37\frac{1}{2}\%$ of 12 **c** $37\frac{1}{2}\%$ of 30.

10 $62\frac{1}{2}\% = \frac{5}{8}$

Use this fact to work out:

a $62\frac{1}{2}\%$ of 6 **b** $62\frac{1}{2}\%$ of 11 **c** $62\frac{1}{2}\%$ of 28.

12.3 Dividing with integers and fractions

Learning objectives

- To divide a fraction or a mixed number by an integer
- To divide an integer by a unit fraction

In this section you will learn about division with a **unit fraction** and an integer.

A unit fraction is a fraction with a numerator of 1.

Look at these two examples.

- $\frac{1}{2} \div 3$
- $3 \div \frac{1}{2}$

They are not the same.

$\frac{1}{2} \div 3$ means work out a third of $\frac{1}{2}$.

The answer is $\frac{1}{6}$ because $\frac{1}{6} \times 3 = \frac{3}{6} = \frac{1}{2}$.

You can also write $\frac{1}{2} \div 3$ as $\frac{1}{2 \times 3} = \frac{1}{6}$.

Just multiply the denominator by 3.

$3 \div \frac{1}{2}$ means work out how many halves make 3.

The answer is 6 because $\frac{1}{2} \times 6 = 3$.

You can write $3 \div \frac{1}{2} = 6$.

Just multiply 3 by the denominator.

Example 5

Work these out. **a** $\frac{1}{4} \div 2$ **b** $\frac{4}{5} \div 6$ **c** $2\frac{1}{2} \div 3$ **d** $8 \div \frac{1}{5}$

a $\frac{1}{4} \div 2 = \frac{1}{4 \times 2} = \frac{1}{8}$ Check that $\frac{1}{8} \times 2 = \frac{1}{4}$.

b $\frac{4}{5} \div 6 = \frac{4}{5 \times 6} = \frac{4}{30} = \frac{2}{15}$ Check that $\frac{2}{15} \times 6 = \frac{4}{5}$.

c $2\frac{1}{2} \div 3 = \frac{5}{2} \div 3$ Write $2\frac{1}{2}$ as an improper fraction.

    $= \frac{5}{2 \times 3} = \frac{5}{6}$ Multiply the denominator by 3.

d $8 \div \frac{1}{5} = 8 \times 5 = 40$ Check that $\frac{1}{5} \times 40 = 8$.

Exercise 12C

1 Work these out.

   a $\frac{1}{2} \div 2$ **b** $\frac{1}{2} \div 4$ **c** $\frac{1}{2} \div 6$ **d** $\frac{1}{2} \div 10$

2 Work these out.

   a $\frac{2}{3} \div 2$ **b** $\frac{2}{3} \div 3$ **c** $\frac{2}{3} \div 4$ **d** $\frac{2}{3} \div 6$

3 Work these out.

   a $\frac{3}{8} \div 2$ **b** $\frac{3}{4} \div 5$ **c** $\frac{5}{12} \div 4$ **d** $\frac{1}{6} \div 4$

   e $\frac{7}{8} \div 2$ **f** $\frac{6}{7} \div 3$ **g** $\frac{2}{9} \div 4$ **h** $\frac{4}{5} \div 8$

4 Work these out.

   a $1\frac{1}{2} \div 2$ **b** $2\frac{1}{2} \div 2$ **c** $3\frac{1}{2} \div 2$ **d** $3\frac{1}{4} \div 2$

5 Work these out.

   a $2\frac{1}{2} \div 4$ **b** $2\frac{1}{3} \div 3$ **c** $4\frac{1}{2} \div 3$ **d** $5\frac{1}{2} \div 4$

   e $3\frac{1}{2} \div 5$ **f** $6\frac{1}{2} \div 4$ **g** $8\frac{1}{2} \div 5$ **h** $2\frac{3}{4} \div 3$

6 Work these out.

   a $7\frac{1}{2} \div 4$ **b** $7\frac{1}{2} \div 5$ **c** $7\frac{1}{2} \div 8$ **d** $7\frac{1}{2} \div 10$

(PS) **7** The perimeter of a square is $13\frac{1}{2}$ cm. Work out the length of each side.

(PS) **8** The perimeter of a regular hexagon is $21\frac{1}{2}$ cm. Work out the length of each side.

9 Work these out.

   a $2 \div \frac{1}{4}$ **b** $2 \div \frac{1}{5}$ **c** $3 \div \frac{1}{3}$ **d** $4 \div \frac{1}{2}$

10 Work these out.

   a $8 \div \frac{1}{2}$ **b** $8 \div \frac{1}{3}$ **c** $8 \div \frac{1}{5}$ **d** $8 \div \frac{1}{8}$

11 Work these out.

    a $4 \div \frac{1}{5}$ **b** $5 \div \frac{1}{4}$ **c** $10 \div \frac{1}{2}$ **d** $2 \div \frac{1}{10}$

12 $4 \div f = 12$

Work out the value of the fraction f.

(PS) **13** A snail, travelling as fast as it can, may move at $\frac{1}{3}$ cm per second. How long does a fast snail take to travel 30 cm?

Challenge: Cycle race

Cycle races take place in a velodrome.

Older tracks have a length of $\frac{1}{3}$ km but modern tracks are $\frac{1}{4}$ km long.

A Work out the number of laps that cyclists would have to make in a 5 km race:

    a on an older track   **b** on a modern track.

B In 1996 Chris Boardman set a one-hour cycling record of 56 km. How many laps is that:

    a on an older track   **b** on a modern track?

12.4 Multiplication with large and small numbers

Learning objective

- To multiply with combinations of large and small numbers mentally

Look at these multiplications.

- 300×2
- 0.3×20
- 0.03×0.2
- 30×0.002

All the answers involve the simple multiplication $3 \times 2 = 6$.

The problem is where to put the decimal point.

Here are the answers.

- $300 \times 2 = 600$
- $0.3 \times 20 = 6$
- $0.03 \times 0.2 = 0.006$
- $30 \times 0.002 = 0.06$

The first one is straightforward. The last one takes more thought.

You need to be able to do multiplications like this *without* using a calculator.

The examples show you how to do this.

Example 6

Work these out.　　　**a** 4 × 500　　　**b** 300 × 60

　a 4 × 5 = 20　　　　　　First just multiply the non-zero digits.

　　4 × 500 = 2000　　　　Because 500 is 5 × 100 you multiply 20 by 100.

　b 3 × 6 = 18　　　　　　First just multiply the non-zero digits.

　　300 × 60 = 18 000　　　300 = 3 × 100 and 60 = 6 × 10 so multiply by 1000.

Example 7

Work these out.　　　**a** 0.002 × 9　　　**b** 0.4 × 0.5

　a 2 × 9 = 18　　　　　　First just multiply the non-zero digits.

　　0.002 × 9 = 0.018　　　0.002 = 2 ÷ 1000 so divide 18 by 1000.

　b 4 × 5 = 20　　　　　　First just multiply the non-zero digits.

　　0.4 × 5 = 2　　　　　　0.4 = 4 ÷ 10 so divide 20 by 10.

　　0.4 × 0.5 = 0.2　　　　0.5 = 5 ÷ 10 so divide 2 by 10.

Example 8

Work these out.　　　**a** 500 × 0.06　　　**b** 0.003 × 70

　a 5 × 6 = 24　　　　　　First just multiply the non-zero digits.

　　500 × 6 = 2400　　　　500 = 5 × 100 so multiply 24 by 100.

　　500 × 0.06 = 24　　　　0.06 = 6 ÷ 100 so divide 2400 by 100.

　b 3 × 7 = 21　　　　　　First just multiply the non-zero digits.

　　0.003 × 7 = 0.021　　　0.003 = 3 ÷ 1000 so divide 21 by 1000.

　　0.003 × 70 = 0.21　　　70 = 7 × 10 so multiply 0.021 by 10.

Exercise 12D

1　Work these out.

　　a　300 × 4　　　**b**　30 × 40　　　**c**　3 × 4000　　　**d**　300 × 400

2　Work these out.

　　a　50 × 80　　　**b**　700 × 6　　　**c**　80 × 900　　　**d**　90 × 800

3 Work these out.

a 0.4×8 b 0.3×0.3 c 0.7×0.05 d 0.09×5

e 0.02×0.4 f 0.06×0.6 g 0.09×0.08 h 0.11×0.6

4 Work these out.

a 20^2 b 40^2 c 60^2 d 300^2

e 0.3^2 f 0.9^2 g 0.02^2 h 0.08^2

5 A rectangular piece of card is 20 cm wide and 30 cm long.

    a   Work out the area of the card. Give your answer in square centimetres (cm²).

    b   Work out the lengths of the sides of the piece of card, in millimetres.

    c   Work out the area of the card, giving your answer in square millimetres (mm²).

    d   Work out the lengths of the sides of the piece of card, in metres.

    e   Work out the area of the card, giving your answer in square metres (m²).

6 Work these out.

a 0.3×6 b 0.4×60 c 30×0.9 d 500×0.4

e 0.08×800 f 0.11×60 g 7000×0.09 h 0.005×300

7 Work these out.

a 0.2×60 b 0.02×6000 c 20×0.06 d 2000×0.006

(PS) 8 A strip of metal is 6 m long and 8 cm wide.

    a   Work out the area in cm².   b   Work out the area in m².

9 $18 \times 35 = 630$

Use this fact to work these out.

a 1.8×35 b 18×0.35 c 180×350 d 0.18×0.35

(PS) 10 $23^2 = 529$

Use this fact to work these out.

a 230^2 b 2.3^2 c 0.23^2

11 Work these out.

a $2 \times 3 \times 4$ b $2 \times 30 \times 4$ c $20 \times 3 \times 40$ d $0.2 \times 3 \times 0.4$

(PS) 12 The lengths of the sides of a large box are 0.4 m, 0.5 m and 0.8 m.

    a   Work out the volume, giving your answer in cubic metres (m³).

    b   Work out the volume, giving your answer in cubic centimetres (cm³).

Challenge: Falling down

If a small object is dropped, the distance it falls in
t seconds is $5t^2$ metres.

A Copy and complete this table to show the distance
fallen in different fractions of a second.

Time (s)	0.1	0.2	0.3	0.4	0.5
Distance (m)			0.45		

B Plot your results on a graph like this.

Join the points with a smooth curve.

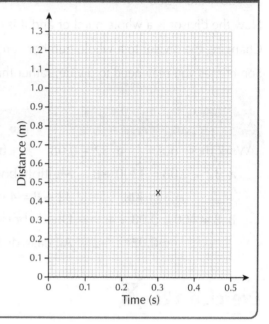

12.5 Division with large and small numbers

Learning objective

- To divide combinations of large and small numbers
  mentally

Key word

divisor

Look at these divisions. They all have the same answer.

- $8 \div 4 = 2$
- $80 \div 40 = 2$
- $800 \div 400 = 2$
- $0.8 \div 0.4 = 2$
- $0.08 \div 0.04 = 2$

It is easy to use multiplication to check that these are correct. For example, in the second one $2 \times 40 = 80$
and in the last one $2 \times 0.04 = 0.08$.

If you start a division that you know is correct, such as $8 \div 2 = 4$, you can see that multiplying or
dividing both numbers by the same integer will give a division with an identical answer.

Suppose you want to work out $30 \div 0.6$.

The number you are dividing by, 0.6, is called the **divisor**.

If you multiply both numbers by 10 you get:

$30 \div 0.6 = 300 \div 6$

Now the divisor is a whole number and it is easy to see that the answer is 50.

Changing the divisor to a whole number can make a division easier to do.

Sometimes you may need to divide, rather than multiply, as shown in the next example.

Example 9

Work these out.   **a** $32 \div 0.08$ **b** $8 \div 400$

a $32 \div 0.08 = 3200 \div 8$ Multiply both numbers by 100 to make the divisor 8.

   $= 400$ $32 \div 8 = 4$ so the answer is 400.

b $8 \div 400 = 0.08 \div 4$ Divide both numbers by 100 to make the divisor 4.

   $= 0.02$ An easy division!

Exercise 12E

1 Work these out.

   a $6 \div 0.2$ **b** $6 \div 0.3$ **c** $12 \div 0.4$ **d** $15 \div 0.5$

   e $20 \div 0.4$ **f** $24 \div 0.8$ **g** $24 \div 0.6$ **h** $35 \div 0.7$

2 Work these out.

   a $30 \div 0.2$ **b** $20 \div 0.4$ **c** $60 \div 0.3$ **d** $120 \div 0.6$

   e $800 \div 0.4$ **f** $360 \div 1.2$ **g** $100 \div 0.2$ **h** $320 \div 1.6$

3 Work these out.

   a $400 \div 20$ **b** $40 \div 20$ **c** $4 \div 20$ **d** $0.4 \div 20$

4 Work these out.

   a $140 \div 20$ **b** $14 \div 20$ **c** $1.4 \div 20$ **d** $14 \div 200$

5 Work these out.

   a $8 \div 20$ **b** $16 \div 80$ **c** $140 \div 200$ **d** $9 \div 300$

   e $80 \div 400$ **f** $12 \div 300$ **g** $33 \div 110$ **h** $450 \div 9000$

6 Work these out.

   a $0.6 \div 10$ **b** $0.6 \div 20$ **c** $0.6 \div 30$ **d** $0.6 \div 60$

(MR) **7** Which is the odd one out? Give a reason for your answer

   a $27 \div 0.9$ **b** $270 \div 90$ **c** $2.7 \div 0.09$ **d** $0.27 \div 0.009$

(MR) **8** Which is the odd one out? Give a reason for your answer

   a $0.48 \div 1.2$ **b** $0.24 \div 0.6$ **c** $0.16 \div 0.4$ **d** $0.8 \div 0.2$

9 Work these out.

   a $12 \div 0.4$ **b** $1.2 \div 40$ **c** $120 \div 400$ **d** $120 \div 0.04$

10 Work these out and put them in order of size, smallest first.

    a $21 \div 0.7$ **b** $0.21 \div 0.7$ **c** $21 \div 700$ **d** $0.21 \div 0.07$

11 Work these out.

    a $20 \div 0.5$ **b** $0.3 \div 0.06$ **c** $5.5 \div 0.11$ **d** $1.8 \div 300$

    e $4.2 \div 0.14$ **f** $6 \div 0.15$ **g** $38 \div 1.9$ **h** $4.5 \div 90$

12 The area of this rectangle is A cm².

20 cm

x cm

    a Explain why $x = \frac{A}{20}$.

    b Work out the value of x when $A = 0.4$.

    c Work out the value of x when $A = 1000$.

13 $1001 \div 77 = 13$

Use this fact to work out:

    a $100.1 \div 77$ **b** $10.01 \div 77$ **c** $10.01 \div 7.7$ **d** $100.1 \div 0.77$.

Reasoning: Fraction or decimal?

These decimals can be written as simple unit fractions.

    $0.1 = \frac{1}{10}$ $0.2 = \frac{1}{5}$ $0.5 = \frac{1}{2}$ $0.05 = \frac{1}{20}$

Look at the decimal calculations below.

A Work each one out.

B Then change each of them to a fraction calculation and check that you get the same answer.

The first one has been done for you.

    a $10 \div 0.2 = 100 \div 2 = 50$ or $10 \div \frac{1}{5} = 10 \times 5 = 50$

    b $20 \div 0.5$ **c** $32 \div 0.1$ **d** $6 \div 0.05$ **e** $3 \div 0.2$ **f** $35 \div 0.5$

Ready to progress?

I can add and subtract mixed numbers.
I can multiply a fraction or a mixed number by an integer.
I can divide an integer by a unit fraction or a unit fraction by an integer.
I can multiply simple large or small decimal numbers without using a calculator.
I can divide simple large or small decimal numbers without using a calculator.

Review questions

1 Work these out.

   a $2\frac{5}{6} + \frac{1}{4}$ b $2\frac{5}{6} - \frac{1}{4}$ c $\frac{1}{4}$ of $2\frac{5}{6}$

2 Work out the missing number in each case.

   a $4\frac{1}{3} + \ldots = 7$ b $3\frac{3}{4} + \ldots = 7\frac{1}{2}$ c $1\frac{7}{10} + \ldots = 5\frac{1}{2}$

3 a Work out the perimeter of this rectangle.

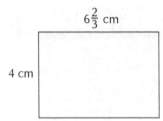

   b Work out the area of the rectangle.

4 a Work out $\frac{1}{5}$ of $3\frac{1}{2}$.   b Increase $3\frac{1}{2}$ by 20%. $20\% = \frac{1}{5}$

5 Work these out.

   a $6 \times \frac{1}{5}$ b $6 \times 1\frac{1}{5}$ c $6 \times 2\frac{2}{5}$ d $2 \times 6\frac{1}{5}$

6 Work these out.

   a 70×70 b 20×5000 c 300^2 d 40×60

7 Work these out.

   a 0.05×400 b 0.8×0.08 c 0.4×0.15 d 90×0.9

8 **a** Calculate the area of this right-angled triangle, giving your answer in square centimetres (cm²).

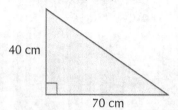

40 cm

70 cm

   b Calculate the area of the triangle, giving your answer in square metres (m²).

9 Work these out.

   a 24 ÷ 0.8   **b** 48 ÷ 0.6   **c** 3.2 ÷ 0.08   **d** 36 ÷ 0.09

10 Work these out.

   a 1200 ÷ 40   **b** 2000 ÷ 50   **c** 8 ÷ 20   **d** 24 ÷ 400

 11 45 × 64 = 2880

Use this fact to work these out.

   a 450 × 640   **b** 4.5 × 6.4

 12 Work out the mean of $2\frac{1}{2}$, $\frac{2}{3}$ and $1\frac{5}{6}$.

13 **a** Write 2.2 as a mixed number.

   b Work out $\frac{1}{3}$ of 2.2.

   Give your answer as a fraction.

   c Work out $3\frac{1}{3}$ + 2.2.

   Give your answer as a mixed number.

14 42 × 75 = 3150

Use this fact to work these out.

   a 3150 ÷ 42   **b** 3150 ÷ 75   **c** 315 ÷ 75   **d** 31.5 ÷ 4.2

Challenge
Guesstimates

You could work with a partner on these questions.

Do not use a calculator.

Sometimes you want to make an estimate of something, even when you do not have exact values to work with. You can still get a rough idea of the size of the answer.

Here is an example. Fill in the gaps to complete it.

1 How many people could stand on a football pitch?
First, think what information you need to know, to answer this.

You need to know the size of the pitch and the amount of room one person takes up.

Pitches vary in size but a typical pitch is about 100 m long and 50 m wide.

> **Hint** Using the numbers 100 and 50 is more sensible than, say, 105 and 55 for making an estimate.

How much room does one person take up? That varies too, but a rectangle measuring about 50 cm by 40 cm seems reasonable.

> **Hint** A rectangle is an easy shape to deal with.

Be careful to use the same units.

The area of the pitch is $100 \times 50 = \ldots$ m²

The area occupied by one person is $0.5 \times 0.4 = \ldots$ m²

So the number of people is $\ldots \div \ldots = \ldots$

This is a very rough estimate – a guesstimate!

Now try these. In each case there are some useful hints. Show how you arrive at your guesstimate each time. Compare your guesstimates with other people's.

2 A pile of sheets of paper stretches from the floor to the ceiling. Estimate the number of sheets.

> **Hint**
> - A4 paper is usually sold in packs of 500 sheets. How thick is one of those packs?
> - The normal height of a door is 2 metres.

3 What is the total mass of all the pupils in your school?

> **Hint**
> - Roughly how many pupils are there in your school?
> - What is the mass of an average pupil?

4 If you left a tap running in your classroom, how long would it take to fill the room with water? Assume all the doors and windows are watertight!

> **Hint**
> - A cold-water tap flows at about 10 to 20 litres per minute.
> - 1000 litres = 1 m³

5 How many books are there in your school library?

> **Hint**
> - On average, how many books are there on one shelf (or on a metre of shelving)?
> - How many shelves (or metres of shelving) are there in the library?

6 If a lift was installed, to take you from the Earth to the Moon, how long would it take for you to get there?

> **Hint**
> - A website says the average distance from the Earth to the Moon is 384 400 km. If you use this number, round it sensibly first.
> - The Guinness Book of Records says that the world's fastest lift in a building has a top speed of 1010 metres per minute.

7 How many cars drive into your school in a year? Count each particular car every time it enters the school.

> **Hint**
> - Who drives in each day?
> - How many days is the school open?

8 How many hairs are there on the head of an average pupil?

> **Hint**
> - Assume the hairs grow 1 mm apart. How many will there be in 1 cm²?
> - What is the area of your scalp? (Imagine it flattened out into a rectangle.)

9 How many words does your mathematics teacher speak in your class in a year?

> **Hint**
> - How many words does a teacher speak in one minute?
> - How many minutes does your teacher speak in a lesson?

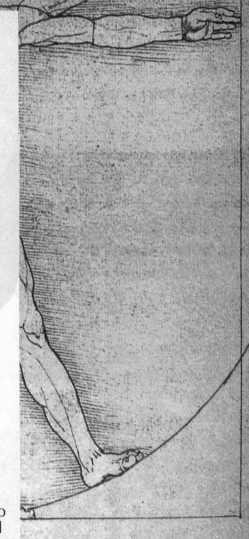

13

Proportion

This chapter is going to show you:

- how to solve problems involving direct proportion
- graphical and algebraic representations of direct proportion
- how to solve problems involving inverse proportion
- graphical and algebraic representations of inverse proportion.

You should already know:

- how to write a simple formula algebraically
- what a ratio is
- how to draw graphs.

About this chapter

The proportions of the parts of the human body change with age.

For an average baby, the length of the head is about a quarter of the length of the whole body.

For an average adult, the length of the head is about an eighth of the length of the whole body.

Ratios of different parts of the body, such as the length of legs compared to the whole body, vary from person to person. Scientists have carried out research to try to find the body ratios that people find most attractive. They have found that the 'ideal' body shape varies from one country to another.

Leonardo da Vinci thought that the ideal proportions should fit in a circle. He made a drawing, called *Vitruvian Man*, to show this.

13.1 Direct proportion

Learning objectives

- To understand the meaning of direct proportion
- To find missing values in problems involving proportion

Key words

direct proportion	proportional
variable	

When you buy petrol you pay a fixed price for each litre. If you buy twice or three times as much as you did last time, you pay twice or three times as much as you paid before. This is an example of **direct proportion**.

Two **variables** (such as the number of litres and the total price you pay) are in direct proportion if, when you multiply one by a number (such as 2 or 3 or 0.5) you need to multiply the other by the same number.

These are some more pairs of variables that are in direct proportion.

- The distance travelled by a car moving at 100 km/hour and the time taken
- The volume of water flowing out of a tap and the time, in seconds, for which the water flows
- The volume of a fizzy drink and the amount of sweetener in it
- The mass of some loose carrots bought in a shop and the cost, in pounds
- The time a light has been on and the cost of the electricity used
- The length of a journey, in miles, and the length of the same journey, in kilometres

Example 1

At a petrol station, 15 litres of petrol cost £20.40.

Work out the cost of: **a** 30 litres **b** 60 litres **c** 5 litres.

a It helps to put the numbers in a table.

Petrol (litres)	15	30	60	5
Cost (£)	20.40			

$30 = 15 \times 2$ The number of litres (15) is multiplied by 2, so the cost (£20.40) is also multiplied by 2.

The cost of 30 litres is £20.40 × 2 = £40.80.

b $60 = 15 \times 4$ The number of litres is multiplied by 4.

The cost of 60 litres is £20.40 × 4 = £81.60.

c $5 = 15 \div 3$ The number of litres is divided by 3. Do the same to the cost.

The cost of 5 litres is £20.40 ÷ 3 = £6.80.

 Hint If you multiply or divide the number of litres by any number, you must multiply or divide the cost by the same number.

You say that the number of litres and the cost in pounds are directly **proportional**.

You can leave out the word 'directly' and just say they are proportional.

You will learn about another sort of proportion later in this chapter.

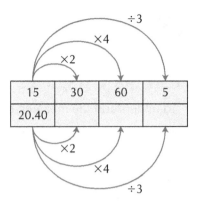

Exercise 13A

1 A train is travelling at a constant speed.

The distance travelled is proportional to the time taken.

In 5 minutes the train travels 13 kilometres.

Copy and complete this table.

Time taken (minutes)	5	10	20	30	45
Distance (km)	13				

2 (FS) Jacob buys 300 g of carrots and they cost 84 pence in total.

Work out the cost of:

a 600 g of carrots **b** 900 g of carrots **c** 150 g of carrots **d** 100 g of carrots.

3 250 ml of cola contains 27 g of sugar.

Work out the amount of sugar in:

a 500 ml of cola **b** 1 litre of cola **c** 2 litres of cola **d** 125 ml of cola.

4 Paulo knows that 1.5 kg of flour is enough to make four small loaves.

a How much flour will he need, to make 16 small loaves?

b How many small loaves can he make from 9 kg of flour?

5 A distance of 5 miles is approximately the same as 8 kilometres (km).

Copy and complete this table to show equivalent distances.

Miles	5	40	100			
Kilometres	8			24	40	200

6 The perimeter of a circle is called the circumference.

The circumference of a circle is proportional to the diameter of the circle.

A circle with a diameter of 3.5 m has a circumference of 11 m.

Work out the circumference of a circle with a diameter of:

a 7 m   **b** 10.5 m   **c** 35 m   **d** 1.75 m.

7 Water is dripping from a tap at a steady rate.

In 15 minutes there are 80 drips.

a Work out the number of drips in one hour.

b Work out the time taken for 800 drips from the tap.

FS **8** In a shop, 100 g of sweets cost 64 pence.

    a Work out the cost of:

      i 300 g of sweets   **ii** 500 g of sweets   **iii** 25 g of sweets.

    b What mass of sweets can you buy for:

      i £1.28   **ii** £6.40   **iii** 32p?

FS **9** The exchange rate between pounds (£) and US dollars (US$) is £41 = US$63.

Copy and complete this table.

Pounds (£)	41	205		328	
US dollars (US$)	63		126		630

10 The pressure of a car tyre can be measured in two different units, bars or psi (pounds per square inch).

A pressure of 2.1 bars is the same as 30 psi.

Copy and complete this table to show conversions between the two units.

Bar	2.1			8.4	12.6
psi	30	10	20		

11 The mass of a steel cable is proportional to its length.

A five-metre length of a particular cable has a mass of 8.2 kg.

    a Work out the mass of 20 metres of the cable.

    b Another length of cable of the same type has a mass of 49.2 kg. How long is it?

FS **12** This table shows the exchange rate between pounds (£) and New Zealand dollars (NZ$).

Pounds (£)	50	150
New Zealand dollars (NZ$)	96	

    a Work out the missing value.

    b Work out the ratio of the two amounts of pounds.

    c Work out the ratio of the two amounts of dollars.

MR **13** Energy content on food labels is given in two different units, kilocalories (kcal) and kilojoules (kJ).

Here is part of a conversion table.

Kilocalories (kcal)	38	
Kilojoules (kJ)	160	800

    a Work out the missing value.

    b Show that the ratio of the two amounts of kilojoules is the same as the ratio of the two amounts of kilocalories.

MR **14** Temperature can be measured in degrees Celsius (°C) or degrees Fahrenheit (°F). Here is a table of values.

Degrees Celsius (°C)	20	30	50	100
Degrees Fahrenheit (°F)	68	86	122	212

Is temperature in degrees Celsius proportional to temperature in degrees Fahrenheit? Give a reason for your answer.

Investigation: Age, height and mass

This table is from a US website.

It shows the average height and the average mass of a boy at different ages.

Because it is a US website, the heights are in inches and the masses are in pounds (lb).

Age (years)	2	4	6	8
Average height (inches)	31	37	42	45
Average mass (lb)	28.4	36.0	46.2	57.2

A Is the average height proportional to age? Use numbers from the table to justify your answer.

B Is the average mass proportional to age? Use numbers from the table to justify your answer.

C Is the average mass proportional to the average height? Use numbers from the table to justify your answer.

13.2 Graphs and direct proportion

Learning objective

- To represent direct proportion graphically and algebraically

Key words

formula   graph

This table shows the relationship between distances measured in miles and in kilometres.

They are in direct proportion.

Distance (miles)	20	30	50	60	70	100
Distance (kilometres)	32	48	80	96	112	160

You can plot these values on a **graph** and join them with a line.

There are two things that you should notice.

- The points are in a straight line.
- The line passes through the origin.

A graph of values of two variables that are in direct proportion always has those properties.

Look back at the pairs of values in the table at the start of this section.

Check that:

- $20 \times 1.6 = 32$ $30 \times 1.6 = 48$ $50 \times 1.6 = 80$

and so on.

If you know that x miles is the same distance as y kilometres, you can write this as a **formula**:

$$y = 1.6x$$

If two variables, x and y, are in direct proportion you can always write a formula:

$$y = mx$$

where m is a number.

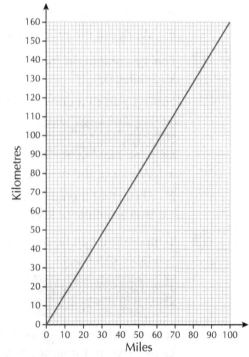

Example 2

Ribbon is sold by the metre. Lucy buys 6 metres and the total cost is £5.04.

a Find a formula for the cost, y pence, of x metres of ribbon.

b Draw a graph to show the cost of different lengths of ribbon.

a The cost is proportional to the length of ribbon, so the formula is $y = mx$.

You need to work out the value of m.

You know that when x is 6 then y is 504.   Notice that y is the cost, in pence.

$504 = m \times 6$

$m = 504 \div 6 = 84$

The formula is $y = 84x$.

b Use the formula to find the costs of different lengths.

Multiply the length by 84 to find the cost.

Choose some values for the length.

Length (x metres)	1	2	3	5	7	10
Cost (y pence)	84	168	252	420	588	840

Plot the points on a graph, then join them up.

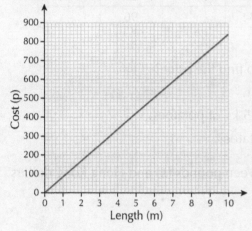

The points should be in a straight line. The line should go to the origin.

Exercise 13B

 1 The perimeter (y cm) of a square, of side x cm, is given by the formula $y = 4x$.

a Copy and complete this table to show values of x and y.

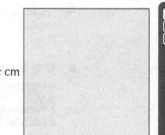

x cm

Side (x cm)	2	5	7	8	10
Perimeter (y cm)					

b Draw a graph to show the values in your table. Label the axes.

FS **2** The cost, £y, of x litres of petrol is given by the formula $y = 1.3x$.

a   Show that 10 litres cost £13.

b   Copy and complete this table.

Petrol (x litres)	10	20	25	30	40
Cost (£y)					

c   Draw a graph to show the figures in your table.

3   A 200 ml glass of a fizzy drink contains 20 g of sugar.

a   Copy and complete this table.

Drink (x ml)	200	100	500	1000
Sugar (y g)	20			

b   You are told that x ml of fizzy drink contain y g of sugar. Show that the formula is $y = 0.1x$.

c   Draw a graph to show the figures in your table.

4   This table shows the price of different masses of potatoes.

Mass (x kg)	0.5	1	1.5	2	3
Price (y pence)		48		96	

a   Work out the missing values of y.

b   What do you multiply the mass by, to find the price?

c   Work out a formula for y in terms of x.

d   Use the formula to find the cost of 7.5 kg of potatoes.

e   Draw a graph to show the cost of potatoes.

5   This graph shows the exchange rate between pounds (£) and Hong Kong dollars (HK$).

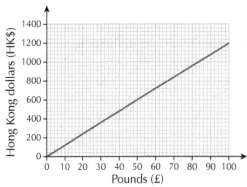

a   Use the graph to complete this table.

Pounds (£x)	25	50	75	100
Hong Kong dollars (HK$y)				

b   Work out a formula for y in terms of x.

c   Use the formula to change £1270 into Hong Kong dollars.

6 A car is travelling at 80 kilometres per hour (km/h).

This graph shows the amounts of petrol used for different distances travelled.

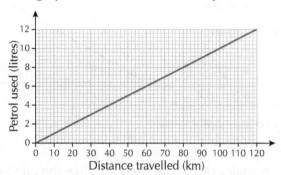

a Use the graph to complete this table.

Distance (x km)	30	60	70	85	110
Petrol (y litres)					

b Write a formula for y in terms of x.

c Use your formula to work out the amount of petrol used to travel 430 km.

7 The speed of a car can be measured in metres per second (m/s) or in kilometres per hour (km/h).

A speed of 5 m/s is the same as a speed of 18 km/h.

a Write 10 metres per seconds in kilometres per hour.

b What do you multiply a speed in metres per second by, to find the speed in kilometres per hour?

c A speed of y km/h is the same as a speed of x m/s.

Write down a formula for y in terms of x.

8 The lengths of the side and the diagonal of a square are proportional.

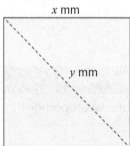

a This table shows possible values of x and y. Fill in the missing values.

Side (x mm)	5	10	15	20	25
Diagonal (y mm)	7				

b Work out a formula for y in terms of x.

c Use your formula to calculate the diagonal of a square, if the length of the side is:

    i 12 mm   **ii** 19 mm   **iii** 31 mm.

9 The angles of this triangle are 30°, 60° and 90°.

The lengths of AB and AC are proportional.

This table shows some possible values of x and y.

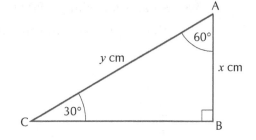

AB (x cm)	4.1	5.2	12.9		
AC (y cm)	8.2			18.8	6.4

a Write down a formula for y in terms of x.

b Work out the missing values in the table.

c Draw your own triangle with angles of 30°, 60° and 90°, like the one in the diagram. Measure x and y and check that they agree with your formula.

Financial skills: Exchange rates

This graph shows the exchange rate between two currencies, the Indian rupee and the Japanese yen.

A Work out a formula connecting the values of the two currencies. Show your method.

B Explain how to use your formula to convert Indian rupees into Japanese yen.

C Could you use your formula to convert Japanese yen into Indian rupees? If so, explain how to do this.

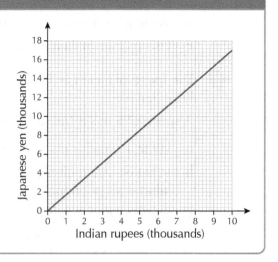

13.3 Inverse proportion

Learning objectives

- To understand what inverse proportion is
- To use graphical and algebraic representations of inverse proportion

Key word

inverse proportion

Suppose you go on a car journey of 120 kilometres.

The time it takes depends on the speed of the car.

- If the car travels at 60 kilometres per hour (km/h) the journey will take 120 ÷ 60 = 2 hours.
- If the car travels at 40 km/h the journey will take 120 ÷ 40 = 3 hours.

This table shows the journey times for a car travelling at various speeds.

Speed (x km/h)	20	30	40	50	60
Time (y hours)	6	4	3	2.4	2

It you multiply the speed by any number, you must divide the time by the same number.

For example:

- $20 \times 2 = 40$ and $6 \div 2 = 3$
- $40 \times 1.5 = 60$ and $3 \div 1.5 = 2$.

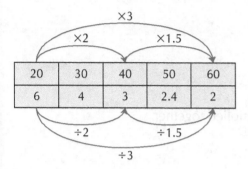

The speed (x) and time (y) are in **inverse proportion**. When you multiply one by a number you divide the other by the *same* number.

In the example above, if you multiply the speed by the time the answer is always 120, because:

speed × time = distance

If the speed is x km/h and the time is y hours, you can write this as a formula:

$$xy = 120$$

When x and y are in inverse proportion you can always write a formula:

$$xy = k$$

where k is a number.

This graph shows the numbers in the table at the start of this section.

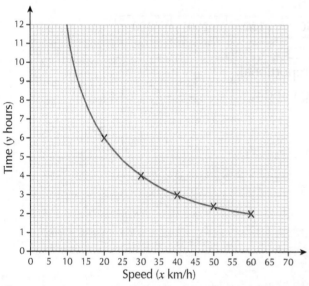

The points are not in a straight line. A smooth curve has been drawn through them.

You can read information from the graph in the same way as you can from a straight-line graph. For example, the curve passes through (10, 12). That tells you that, at a speed of 10 km/h, it would take 12 hours to make the journey.

Graphs of inverse proportion are always this shape.

The variables do not have to be x and y. You can use any letters.

Example 3

The area of a rectangular field is 2400 m².

The sides of the field are *a* metres and *b* metres long.

This table shows possible lengths for the sides of the field.

a metres	40	50	
b metres			80

a Calculate the missing values.

b Work out a formula connecting *a* and *b*.

  a The area of the field is the lengths of the two sides multiplied together.

  When $a = 40$: $40 \times b = 2400 \rightarrow b = 2400 \div 40 = 60$

  When $a = 50$: $50 \times b = 2400 \rightarrow b = 2400 \div 50 = 48$

  When $b = 80$: $a \times 80 = 2400 \rightarrow a = 2400 \div 80 = 30$

  b The formula is $ab = 2400$.

  In this case the variables, *a* and *b*, are inversely proportional.

Exercise 13C

1 A train is travelling a distance of 600 km.

   a How long does the train take if it travels at a speed of 100 km/h?

   b If the train travels at 150 km/h, how long will it take for the same journey?

   c Copy and complete this table.

Speed (*x* km/h)	100	150	120	200	300
Time (*y* hours)					

   d Show that *x* and *y* are inversely proportional.

   e Write down a formula connecting *x* and *y*.

   f Draw a pair of axes like this.

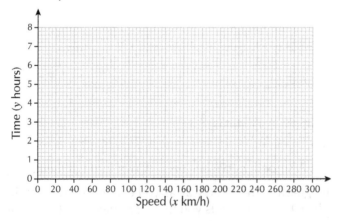

   g Plot the points from the table in part **c** and join them with a smooth curve.

2   A teacher has £1000 to spend on books.

   a   Some books cost £10 each. How many can the teacher buy?

   b   Other books cost £5 each. How many can the teacher buy?

   c   Copy and complete this table.

Cost of a book (£x)	2	2.50	5	10	20	25
Number bought (y)						

   d   Show that x and y are inversely proportional.

   e   Write down a formula connecting x and y.

   f   Draw a pair of axes like this.

   g   Draw a graph to show the information in the table.

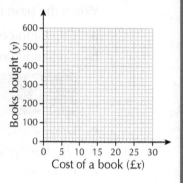

3   The graph shows the time taken by an aeroplane to travel between two airports, at different speeds.

   a   Use the graph to find the time taken when the aeroplane flies at a speed of 800 km/h.

   b   Use the graph the find the speed when the journey takes 4 hours.

   c   The time (y hours) is inversely proportional to the speed (x km/h).

      Use your answers to parts **a** and **b** to find a formula connecting x and y.

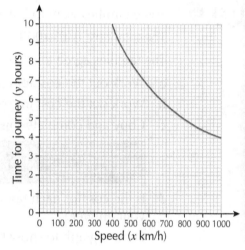

4   Some children are measuring the lengths of their paces and how many paces they take to walk 12 metres.

   a   If the length of a pace is 0.5 metres, how many paces will they take to walk 12 metres?

   b   Copy this table and fill in the missing values.

Length of pace (p metres)	0.5	0.6	1	1.2
Number of paces (n)				

   c   Show that p and n are inversely proportional.

   d   Write down a formula connecting p and n.

MR **5** An isosceles triangle has an area of 100 cm².

The length of the base (*b* cm) is inversely proportional to the height (*h* cm).

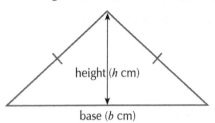

height (*h* cm)

base (*b* cm)

When the base is 20 cm then the height is 10 cm.

a Copy and complete this formula: $bh = \ldots$.

b Copy and complete this table to show possible values of *b* and *h*.

Base (*b* cm)	20	16	12.5	10	8
Height (*h* cm)	10				

c Draw a graph to show how the height varies with the base.

d Use your graph to find the height, when the base is 15 cm.

e Use your formula to check the answer to part **d**.

FS **6** Some families want to buy play equipment for their local park.

The total cost is £30 000.

The families agree to share the cost between them, equally.

a Work out the cost for each family, when there are 20 families.

b Work out the cost for each family, when there are 30 families.

c Copy and complete this table of values.

Number of families (*n*)	10	20	30	40	50	60
Cost for each family (£*c*)						

d Is the cost for each family inversely proportional to the number of families? Justify your answer.

e Draw a graph to show how the cost varies with the number of families.

f They decide that each family should not pay more than £800. Use your graph to work out the smallest number of families that need to take part.

g Write down a formula connecting *n* and *c*. Use it to check your answer to part **f**.

Activity: Different rectangles, same area

A Draw two different rectangles, each with an area of 48 cm².

B Copy and complete this table to show possible lengths for the height and base of a rectangle with an area of 48 cm².

Base (*x* cm)	4	5	6	8	10	12
Height (*y* cm)						

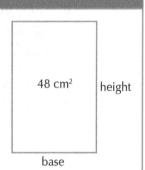

48 cm²   height

base

C Draw a graph to show your values.

D Write down a formula to show the connection between *x* and *y*.

E Use your graph to estimate the side of a square with an area of 48 cm².

13.4 Comparing direct proportion and inverse proportion

Learning objective

- To recognise direct and inverse proportion and work out missing values

Here is a summary of what you have learnt about direct proportion and inverse proportion.

Direct proportion

x	3	4.5	6	15	20
y	24	36	48	120	160

The variables x and y are directly proportional.

If you multiply (*or divide*) a value of x by a number, you must multiply (*or divide*) y by the same number.

You can always write $y = mx$, where m is a number.

In this example, $y = 8x$.

Inverse proportion

x	3	4.5	6	15	20
y	24	16	12	4.8	3.6

The variables x and y are inversely proportional.

If you *multiply* a value of x by a number, you must *divide* y by the same number.

You can always write $xy = k$, where k is a number.

In this example, $xy = 72$.

Check that *multiplying* each pair of numbers does give 72.

Example 4

Here are the values of two variables, p and q.

p	20	50
q	90	

a Find the missing value, when q is directly proportional to p.

b Find the missing value, when q is inversely proportional to p.

  a Look at the values in the first column.

    $q \div p = 90 \div 20 = 4.5$

      $q = 4.5p$ Because they are in direct proportion.

    When $p = 50$ then $q = 4.5 \times 50 = 225$.

    $q = 225$

  b Look at the values in the first column.

    $pq = 20 \times 90 = 1800$

    $pq = 1800$ Because they are in inverse proportion.

    When $pq = 1800$ and $p = 50$ then $50 \times q = 1800 \rightarrow q = 1800 \div 50 = 36$.

1 George is walking at a constant speed.

In 5 minutes he walks 400 metres.

a Is the distance travelled (d metres) proportional to the time taken (t minutes)?

b Write down a formula for d in terms of t.

c Use the formula to work out how far George walks in 8.5 minutes.

2 Anne is doing a sponsored walk of 20 km.

a How long will it take if she walks at 5 km/h?

b How long will it take if she walks at 8 km/h?

c Explain why the time taken (t hours) is inversely proportional to her walking speed (w km/h).

d Write down a formula connecting t and w.

(MR) These are some tables of values. Say whether they show direct proportion, inverse proportion or neither of these.

If the variables are directly or inversely proportional, work out the formula.

a

x	12	17	5	14
y	36	51	20	42

b

c	18	30	12	1.5
d	5	3	7.5	60

c

f	5.6	9.4	63.8	3.6
r	75.6	126.9	861.3	48.6

d

u	12	16	14.8	46.25
w	12	9	10	3.2

(MR) **4** How can you tell that this graph does not show direct proportion?

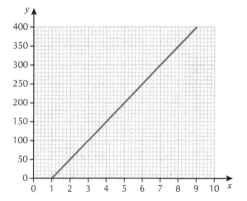

5 **a** Two points on this line are $(4, \ldots)$ and $(7, \ldots)$.

Find the missing y-coordinates.

b Show that x and y are not in inverse proportion.

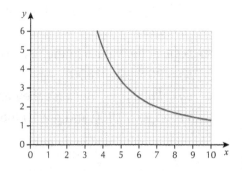

6 **a** Two variables, x and y, are in direct proportion.

When $x = 40$, $y = 10$.

Work out a formula for y in terms of x.

b Two more variables, x and y, are in inverse proportion.

When $x = 40$, $y = 10$.

Work out a formula connecting x and y.

7 This graph shows two variables that are in direct proportion.

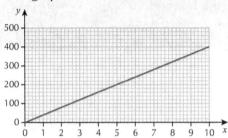

a Write down the coordinates of two points on the line.

b Work out a formula for y in terms of x.

8 This graph shows two variables that are in inverse proportion.

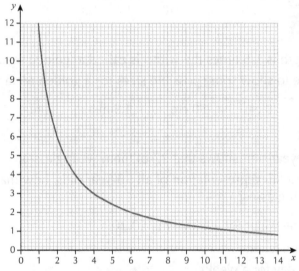

a Write down the coordinates of three points on the line.

b Work out a formula connecting x and y.

Reasoning: Looking for proportion

The perimeter of a rectangle is 20 cm.

A Given that one side of the rectangle is 3 cm, show that the other side is 7 cm.

B Work out three other pairs of possible values for the two sides of the rectangle.

C Draw a graph to show the pairs of values and draw a line through them.

D Are the lengths of the two sides in direct proportion? Justify your answer.

E Are the lengths of the two sides in inverse proportion? Justify your answer.

Ready to progress?

I can recognise and use formulae for two variables that are in direct proportion.
I can recognise and draw a graph for two variables that are in direct proportion.

I can work out a formula connecting two variables that are in direct proportion.
I can work out a formula connecting two variables that are in inverse proportion.
I can recognise and draw a graph for two variables that are in inverse proportion.
I can decide whether two variables are in direct proportion or inverse proportion.

Review questions

1   The number of pages of a particular book and the time it takes to read them are proportional.

Work out the missing numbers in this table.

Number of pages	20	40	100
Time to read them (minutes)	17		

2   A rectangle in which the length is 1.6 times the width is called a 'golden rectangle'.
Copy and complete this table to show some possible sizes for a golden rectangle.

Width	5 cm	20 cm	50 cm	1.5 m	4 m
Length					

3   A survey of one hundred 50-year-olds shows that 27 of them wear glasses.

If the sample in the survey is a fair representation of all 50-year-olds, how many out of 2000 do you expect will wear glasses?

4   The average cost of heating the water for a shower is £0.20.

a   Work out the cost, in pounds, of:

  i   7 showers (one each day for a week)

  ii   30 showers (one each day for a month)

  iii   365 showers (one each day for a year)

b   Write down a formula for the cost (£c) of n showers.

5   The cost of a taxi journey is proportional to the distance travelled.

a   Work out the missing values in this table.

Distance (x km)	4	8	20	2
Cost (£y)	14			

b   Work out a formula for the cost (£y) of travelling x kilometres.

c   Draw a graph to show the figures in the table.

d   Explain why the graph shows that the cost is proportional to the distance travelled.

6 The time taken to complete a rail journey is inversely proportional to the average speed of the train. At a speed of 100 km/h the journey takes 4.5 hours.

Work out the missing values in this table.

Speed (km/h)	100	50	150
Time (hours)	4.5		

7 The time taken to walk to the end of a playing field is inversely proportional to the walking speed. Amelia walks at 1.6 m/s and takes 200 seconds.

a Lily walks at 0.8 m/s. How long does she take?

b Emily runs at 4.8 m/s. How long does she take?

8 These are the lengths and widths of four rectangles with the same area.

Length (x cm)	120	100	80	60
Width (y cm)	10	12	15	20

a Show that the length is inversely proportional to the width.

b The length of a fifth rectangle is 40 cm. Work out its width.

c Work out a formula connecting x and y.

d Use the dimensions of the five rectangles to draw a graph to show values of x and y.

9 A fence is being put along one side of a school playground.

The fence is made from separate panels. The number of panels needed is inversely proportional to the length of each panel.

a Work out the missing numbers in the table.

Length of panel (l m)	1.2	1.5	1.8	2.4	3.0
Number of panels, n	150	120			

b Work out a formula connecting l and n.

10 This graph shows the values of two variables that are in inverse proportion.

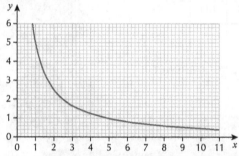

a Find the missing coordinate for each of these points on the line.

i (5,...)   ii (10,...)   iii (2,...)   iv (..., 5)

b Work out a formula connecting x and y.

c Use your formula to find the value of y when $x = 25$.

11 The table shows values of r and s.

a If s is directly proportional to r, work out the missing value of s.

b If s is inversely proportional to r, work out the missing value of s.

r	12	48
s	8	?

Challenge
Planning a trip

A youth leader is planning a coach trip for some youth club members.

1 Cost of the coach

The total cost of the coach is £800.

This must be shared equally by the people on the coach.

a Copy and complete this table.

Number of passengers (x)	50	40	32	25	20
Cost for each person (£y)	16				

b Show that the number of passengers and the cost per person are in inverse proportion.

c Draw a graph to show how the cost per person varies with the number of people.

2 Time and distance

On the motorway the coach travels at a constant speed of 90 km/h.

a How far does the coach travel in 30 minutes on the motorway?

b How far does the coach travel in 10 minutes on the motorway?

c Copy and complete this table.

Time (minutes)	10	20	30	40	50	60
Distance (km)						

d Is the distance proportional to the time? Justify your answer.

e Draw a graph to show the data in the table.

3 Speed and time

The time taken to complete the whole journey is inversely proportional to the average speed.

a Copy and complete this table.

Average speed (x km/h)	50	60	75	80	100
Time (y hours)	4.8	4			

b Write down a formula connecting x and y.

c Draw a graph to show how the time varies with the average speed.

4 Fuel consumption

The amount of fuel the coach uses varies with the distance.

Look at the data in this table. It shows the amounts of fuel used when the coach is travelling on the motorway at 90 km/h.

Fuel used (f litres)	46	82	14	80	100
Distance travelled (d km)	115	205	35		

a Is the relationship between fuel used and distance travelled:

  i direct proportion
  ii inverse proportion
  iii neither of these?

b Use your answer to part a to write a formula for d in terms of f.

The fuel consumption of the coach is measured in litres per kilometre (litres/km). The fuel consumption varies with speed.

Look at the data in this table.

Speed (km/h)	50	60	70	80	100
Fuel consumption (litres/km)	3.2	2.6	2		

c Is the relationship between speed and fuel consumption:

  i direct proportion
  ii inverse proportion
  iii neither of these?

Give a reason for your answer.

14

Circles

This chapter is going to show you:

- the names of the parts of a circle
- how to calculate the circumference of a circle
- how to calculate the area of a circle.

You should already know:

- the words 'radius' and 'diameter'
- how to use compasses to draw a circle
- what a regular polygon is.

About this chapter

Circles are all around us.

The circle is probably the most important shape in the universe. It is also the most mysterious.

You can only measure the circumference (perimeter) of a circle in terms of a number called π. You say this as 'pi'. This number is also mysterious. It cannot be written exactly as a number and its decimal places go on for ever. Its decimal digits never settle into a permanent repeating pattern. They appear to be randomly distributed, although no one has proved this yet.

In the 20th and 21st centuries, mathematicians and computer scientists, helped by increasing computational power, extended the decimal representation of π to over 10 trillion digits. Memorising π is now a sport, with records set by people who can remember more than 67 000 digits.

14.1 The circle and its parts

Learning objective

- To know the definition of a circle and the names of its parts

A circle is a set of points that are all the same distance from a fixed point, called the **centre**. The centre of the circle is usually called O.

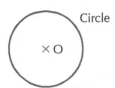
Circle

You must learn all of these words for the different parts of a circle.

- **Circumference**: the length round the outside of the circle. It is a special name for the perimeter of a circle.

Circumference
C

- **Arc**: a part of the circumference of the circle.

Arc

- **Radius**: the distance from the centre of a circle to its circumference. The plural of radius is *radii*.

Radius
r

- **Diameter**: the distance across a circle, through its centre. The diameter d of a circle is twice its radius r, so $d = 2r$.

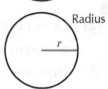
Diameter
d

- **Chord**: a straight line that joins two points on the circumference of a circle. If the chord passes through the centre, it is also a diameter.

Chord

- **Tangent**: a straight line that touches a circle at only one point on its circumference.

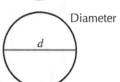

Tangent

- **Segment**: a portion of a circle that is enclosed by a chord and an arc.

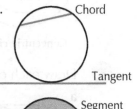
Segment

- **Sector**: a portion of a circle enclosed by two radii and an arc.

Sector

- **Semicircle**: one half of a circle; either of the parts cut off by a diameter.

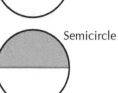
Semicircle

Exercise 14A

1 **i** Measure the radius of each circle, giving your answer in centimetres.

   ii Write down the diameter of each circle.

   a   **b**   **c**

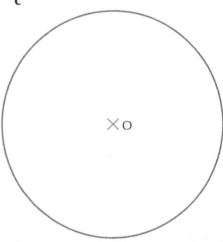

2 Draw circles with these measurements.

   a radius = 2.5 cm **b** radius = 3.6 cm **c** diameter = 8 cm **d** diameter = 6.8 cm

3 Draw each of these shapes accurately. Use a ruler, compasses and a protractor.

   a   **b**   **c**   **d**

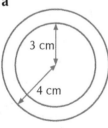

Concentric circles

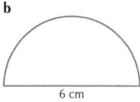

Semicircle

Quadrant of
a circle

Sector of
a circle

4 Draw each of these shapes accurately.

   a   **b**   **c**

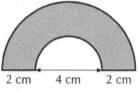

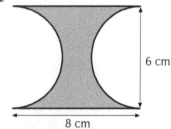

5 Draw a circle with centre O and a radius of 4 cm.

Use a protractor to draw six radii that form angles of 60° at the centre of the circle, as shown in the diagram. Join the six points where the radii meet the circumference, to make a regular hexagon.

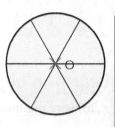

a Explain why the angles between the radii must be 60°.

b Use a similar method to draw:

    i a regular pentagon   **ii** a regular octagon   **iii** a regular decagon.

Activity: Finding the centre of a circle

Draw a circle around a circular object so that you do not know where the centre is.

Draw a chord AB on the circle.

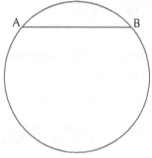

Mark a point X halfway between A and B, as the midpoint of AB.

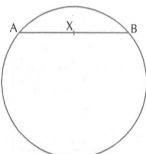

Draw a line that is perpendicular to the chord and that passes through X.

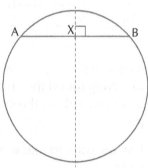

Now repeat these steps for a different chord CD with midpoint Y.

Where the perpendiculars intersect is the centre of the circle, O.

Repeat for different circles.

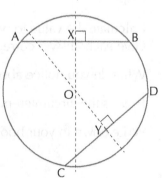

14.2 Circumference of a circle

Learning objective

• To work out the relationship between the circumference and diameter of a circle

How can you measure the circumference of a circle exactly?

Is there a relationship between the length of the diameter and the circumference?

Exercise 14B will show you.

Exercise 14B

You will need compasses, a 30 cm ruler and a piece of fine, high-quality string at least 40 cm long.

Copy this table and then draw circles with the given radii.

Radius r (cm)	Diameter d (cm)	Circumference C (cm)	$C \div d$
1			
1.5			
2			
2.5			
3			
3.5			
4			
4.5			
5			
5.5			
6			

Measure the circumference of each circle by placing the string round the circumference, as shown. Make a pencil mark on the string where it meets its starting point.

Use the ruler to measure this length. Complete the table.

Calculate the value to write in the last column. Give each answer correct to one decimal place.

What do you notice about the numbers in the last column?

How is the circumference related to the diameter?

Write down in your book what you have found out.

Activity: Making nets for cones

A Draw a circle on paper and cut it out.

B Draw a narrow sector on the circle and cut it out.

C Make a cone with the remaining larger sector.

D What happens as you increase the size of the sector you remove?

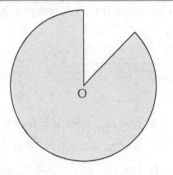

14.3 Formula for the circumference of a circle

Learning objective

• To calculate the circumference of a circle

Key word
π

In Exercise 14B, you should have found that the circumference, C, of a circle with diameter d, is given approximately by the formula $C = 3d$.

In fact, to get a more accurate value for the circumference, you need to multiply the diameter by a number that is slightly larger than 3.

This special number is represented by the Greek letter π (pronounced pi). It is impossible to write down the value of π exactly, as a fraction or as a decimal, so you will use approximate values.

The most common of these are:

• $\pi = 3.14$ (as a decimal rounded to two decimal places)

• $\pi = 3.142$ (as a decimal rounded to three decimal places)

• $\pi = 3.141\ 592\ 654$ (on a scientific calculator)

• $\pi = \frac{22}{7}$ (as a fraction).

Mathematicians have used computers to calculate π to trillions of decimal places. So far, no repeating pattern has ever been found.

Look for the π key on your calculator.

On some calculators, you may need to key in **SHIFT** $\times 10^x$ to use π.

So, the formula for calculating the circumference, C, of a circle with diameter d is written as:

$$C = \pi d$$

As the diameter is twice the radius, r, the circumference is also given by the formula:

$$C = \pi d = \pi \times 2r = 2\pi r$$

Example 1

Calculate the circumference of each circle. Give each answer correct to one decimal place.

a
6 cm

b
3.4 m

a The diameter d = 6 cm, which gives:

$C = \pi d = \pi \times 6 = 18.8$ cm (to 1 dp)

b The radius r = 3.4 m, so d = 6.8 m. This gives:

$C = \pi d = \pi \times 6.8 = 21.4$ m (to 1 dp)

Example 2

Calculate the perimeter, P, of this semicircle. Give your answer correct to one decimal place.

First, calculate the circumference of a circle with d = 8 cm.

$C = \pi d = \pi \times 8 = 25.12$ cm

To work out the length of the curved part of the semicircle, divide the circumference by 2.

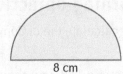
8 cm

$25.12 \div 2 = 12.56$

Now add 8 to work out the total perimeter.   This is the diameter.

So $P = 12.56 + 8 = 20.56 = 20.1$ cm (to 1 dp).

Exercise 14C

In this exercise, take π = 3.14 or use the π key on your calculator.

1 Calculate the circumference of each circle. Give each answer correct to one decimal place.

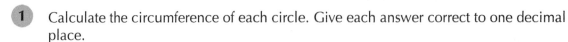

a

7 cm

b

11 mm

c
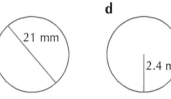
21 mm

d
2.4 m

e

1.4 cm

2 The diameter of a 2p coin is 26 mm.

Calculate the circumference of the coin, giving your answer correct to the nearest millimetre.

3 The London Eye has a diameter of 120 m.

How far would you travel in one complete revolution of the wheel? Give your answer correct to the nearest metre.

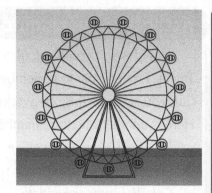

4 The diagram shows the dimensions of a running track at a sports centre. The bends at the ends are semicircles.

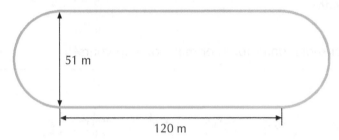

Calculate the distance round the track. Give your answer correct to the nearest metre.

5 The Earth's orbit can be taken to be a circle with a radius of approximately 150 million kilometres.

Calculate the distance the Earth travels in one orbit of the Sun. Give your answer correct to the nearest million kilometres.

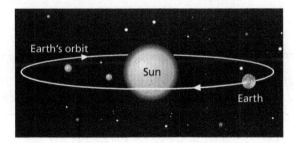

6 By measuring the diameter, calculate the perimeter of this semicircular shape. Give your answer correct to one decimal place.

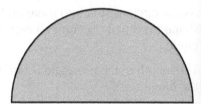

PS **7** The curved parts of this shape are all semicircles.

Calculate the perimeter of the shape. Give your answer correct to one decimal place.

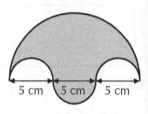

MR **8** The distance round a circular running track is 200 m. Calculate the radius of the track. Give your answer correct to the nearest metre.

14.4 Formula for the area of a circle

Learning objective

• To calculate the area of a circle

This circle has been split into 16 equal sectors. These have been placed together to form a shape that is roughly a rectangle.

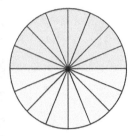 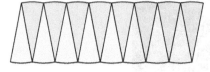 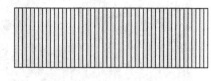

As the circle is split into more and more sectors, and then these are placed together, the resulting shape eventually becomes a rectangle. The area of this rectangle will be the same as the area of the circle.

The length of the rectangle is half the circumference, C, of the circle and its width is the radius, r, of the circle.

$$\tfrac{1}{2}C$$

r ⎢ A

So, the area, A, of the rectangle is given by:

$$A = \tfrac{1}{2} C \times r$$

The formula for the circumference of a circle is $2\pi r$, therefore:

$$A = \tfrac{1}{2} \times 2\pi \, r \times r$$
$$= \pi r \times r$$
$$= \pi r^2$$

Hence, the formula for the area, A, of a circle of radius r is:

$$A = \pi r^2$$

Example 3

Calculate the area of each circle. Give each answer correct to one decimal place.

a

b

a The radius, $r = 3$ cm, which gives:

$$A = \pi r^2 = \pi \times 3^2$$
$$= 9\pi$$
$$= 28.3 \text{ cm}^2 \text{ (to 1 dp)}$$

When using a calculator, you can use the 'square' key .

Simply key in:

b The diameter $d = 3.4$ m, so $r = 1.7$ m. This gives:

$$A = \pi r^2 = \pi \times 1.7^2$$
$$= 2.89\pi$$
$$= 9.1 \text{ m}^2 \text{ (to 1 dp)}$$

Note that you can also leave your answers in terms of π.

For example, this may be necessary when the use of a calculator is not allowed.

This would give answers of 9π and 2.89π for the two examples above.

Example 4

Calculate the area of this semicircle. Give your answer in terms of π.

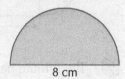

First, calculate the area of a circle with $r = 4$ cm.

$$A = \pi r^2 = \pi \times 16 = 16\pi \text{ cm}^2$$

To work out the area of the semicircle, divide this by 2.

$$A = 16\pi \div 2 = 8\pi \text{ cm}^2$$

In this exercise, take π = 3.14 or use the π key on your calculator.

1 Calculate the area of each circle. Give your answers correct to one decimal place.

a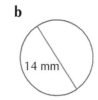
1 cm

b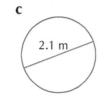
14 mm

c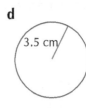
2.1 m

d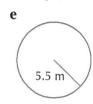
3.5 cm

e
5.5 m

2 Calculate the area of a circular tablemat with a diameter of 21 cm. Give your answer correct to the nearest square centimetre.

3 Measure the diameter of a 1p coin, to the nearest millimetre. Calculate the area of one face of the coin, giving your answer correct to the nearest square millimetre.

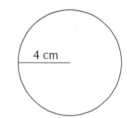

MR **4** Randal is working out the area of this circle.

This is his working.

$$Area = \pi \times d = \pi \times 8 = 8\pi \; cm$$

Explain why Randal's working is wrong.

Write down the correct answer to the problem.

4 cm

5 The diagram represents a sports ground. The bends are semicircles. Calculate the area of the sports ground. Give your answer correct to the nearest square metre.

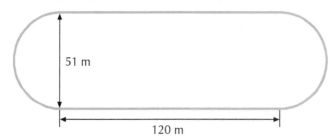
51 m

120 m

6 The minute hand on a clock has a length of 13 cm. Calculate the area swept by the minute hand in one hour. Give your answer in terms of π.

7 Calculate the area of one face of this semicircular protractor. Give your answer correct to one decimal place.

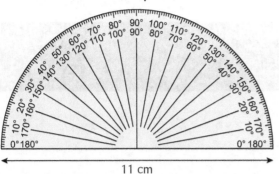

11 cm

(MR) **8** Finlay has cut out a circle of radius 5 cm and Jackson has cut out a circle of radius 10 cm.

Jackson says to Finlay:

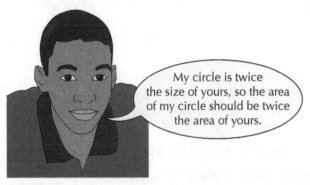

My circle is twice the size of yours, so the area of my circle should be twice the area of yours.

Calculate the areas of both circles, giving your answers in terms of π.

Is Jackson correct? Give a reason for your answer.

Problem solving: Circle problems 🖩

A Calculate the area of each shape. Give your answers correct to one decimal place.

a

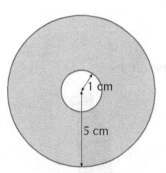

1 cm

5 cm

b

10 cm

10 cm

c

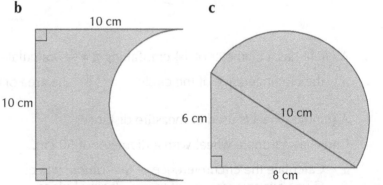

6 cm

10 cm

8 cm

B A circular lawn has an area of 200 m². Calculate the radius of the lawn. Give your answer correct to one decimal place.

C A circular disc has a circumference of 15 cm. Calculate the area of one face of the disc. Give your answer correct to one decimal place.

Ready to progress?

I can use the formula $C = \pi d$ when calculating the circumference of a circle.
I can use the formula $A = \pi r^2$ when calculating the area of a circle.

Review questions

1   This is a sketch of a sector of a circle.

Make an accurate, full-size drawing of this sector. Use a ruler, compasses and a protractor.

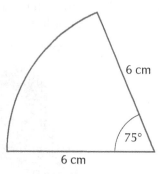

2   Calculate:

i   the circumference   ii   the area of each circle.

Give you answers correct to one decimal place.

a   b   c

9 mm

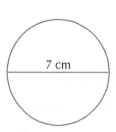

7 cm

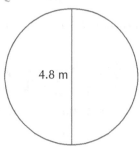

4.8 m

3   A circle has a diameter of 14 cm. Taking $\pi = \frac{22}{7}$, calculate:

a   the circumference of the circle   b   the area of the circle.

4   A trundle wheel is used to measure distances.

Chris has a trundle wheel with a diameter of 50 cm.

a   Calculate the circumference of the trundle wheel. Give your answer correct to one decimal place.

b   Chris uses the trundle wheel to measure the length of a car park.

The trundle wheel rotates 85 times. What is the length of the car park, to the nearest metre?

5 a The cross-section of a cylindrical cotton reel is a circle.

The diameter of the circle is 3 cm.

Calculate the circumference of the circle. Give your answer correct to one decimal place.

b The cotton reel holds 100 metres of cotton thread.

About how many times is the cotton wrapped around the reel?

Round your answer to the nearest ten.

 6 The wheels of a bicycle have a diameter of 26 inches.

You are given that 1 inch = 2.54 cm.

a Convert 26 inches into centimetres. Give your answer correct to the nearest centimetre.

b Calculate the circumference of each wheel. Give your answer correct to the nearest centimetre.

c Lyndon cycles a distance of 10 km.

Work out the number of times the wheels turn. Round your answer to the nearest 10.

 7 The diagram shows a rectangle inside a circle.

The diameter of the circle is 10 cm and the lengths of the sides of the rectangle are 6 cm and 8 cm.

Calculate the area of the shaded part of the circle. Give your answer correct to one decimal place.

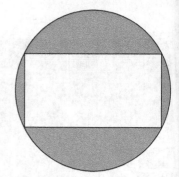

 8 A machine makes sheets of metal. Each sheet measures 1 m by 50 cm.

The sheet is then cut into small squares each measuring 5 cm by 5 cm.

A circular disc with diameter 4 cm is then pressed from each square.

The metal left is melted down and recycled.

a How many squares can be cut from the sheet of metal?

b Write down the area of each small square.

c Calculate the area of a circular disc. Give your answer correct to one decimal place.

d Work out the percentage of metal left from each square. Give your answer correct to one decimal place.

An athletics stadium is being redeveloped.

1 The shot-put circle is going to be resurfaced.

> Give your answers correct to two decimal places.

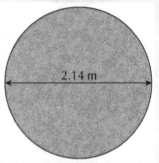

a What is the perimeter of the circle?

b What is the area of the circle?

c The cost for resurfacing is £32 per square metre. How much will it cost to resurface the shot-put circle?

2 The high-jump zone is also going to be resurfaced.

a What is the area of the high-jump zone?

b Resurfacing costs £57.50 per square metre. How much will it cost to resurface the high-jump zone?

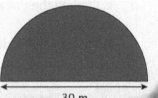

30 m

3 The running track is going to be repainted. This diagram shows the dimensions of the running track.

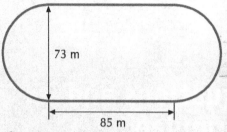

  a What is the total distance around the inside line?

One litre of paint is mixed with 2 litres of water.
The paint is bought in large 20-litre tins.
Each tin costs £52.70.

  b How much water do you need to mix with 50 tins of paint?

  c What is the cost of 50 tins of paint?

4 The area inside the running track is going to be returfed.

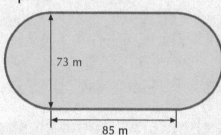

  a Calculate the area of the turf inside of the running track. Give your answer correct to the nearest 100m².

  b A quote for returfing is £3.60 per square metre (including labour costs). How much will it cost to returf the area?

5 This sketch shows the dimensions of the sandpit for the long jump.

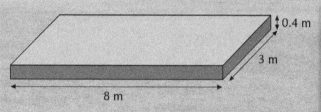

  a How many cubic metres of sand does the sandpit hold when it is full?

  b The mass of 0.6 m³ of sand is 1 tonne. How many tonnes of sand are needed to fill the sandpit?

6 A ramp for wheelchair access into the stadium is going to be built.

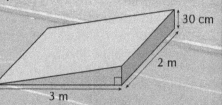

  a What volume of concrete will be needed to make the ramp?

> **Hint** The volume of the triangular prism is half the volume of a cuboid with the same measurements.

  b Concrete costs £48 per cubic metre. How much will the concrete for the ramp cost?

15

Equations and formulae

This chapter is going to show you:

- how to solve equations involving brackets and fractions
- how to solve equations where the variable occurs more than once
- how to rearrange formulae.

You should already know:

- how to solve simple equations
- how to write simple formulae.

About this chapter

Behind most of today's technology lie equations, whether it is a space probe going into orbit or a computer game being designed. When scientists plan a mission, such as taking a spacecraft to Mars or trying to land on a comet, they have to carry out extremely detailed plans for the mission. They use formulae and equations to simulate the path the spacecraft will take, and try out different paths to see which is the best one. They also use equations to see what effect small changes will have on this path and to try to model things that could go wrong. These equations are much more complicated than any that you have come across so far.

In this chapter you will learn how to extend the techniques you have already learned to deal with more complex equations.

15.1 Equations with brackets

Learning objective

Key word

inverse operation

- To solve equations involving brackets

You have learnt how to solve equations by using **inverse operations**. For example:

- the inverse of subtract 2 is add 2
- the inverse of multiply by 3 is divide by 3.

You also know how to multiply out brackets in algebraic expressions.

Example 1

Solve these equations.

a $3x - 2 = 15$ **b** $3(x - 2) = 15$ **c** $\frac{1}{3}(x - 2) = 15$

a $3x - 2 = 15$

$3x = 17$ Add 2 to both sides. $15 + 2 = 17$

$x = \frac{17}{3}$ Divide both sides by 3.

$= 5\frac{2}{3}$ $17 \div 3 = 5$ remainder 2.

Write the answer as a mixed number, because the decimal for $5\frac{2}{3}$ is 5.666 666… and you would need to round it.

b $3(x - 2) = 15$ There are two ways to solve this.

Method 1

$x - 2 = 5$ Divide both sides by 3. $15 \div 3 = 5$

$x = 7$ Add 2 to both sides.

Method 2

$3x - 6 = 15$ Multiply out the brackets. $3(x - 2) = 3x - 6$

$3x = 21$ Add 6 to both sides.

$x = 7$ Divide by 3. $21 \div 3 = 7$

Make sure you can use both of these methods.

c $\frac{1}{3}(x - 2) = 15$ Finding $\frac{1}{3}$ is the same as dividing by 3.

$x - 2 = 45$ Multiply both sides by 3. $15 \times 3 = 45$

$x = 47$ Add 2 to both sides. $45 + 2 = 47$

The equation in part **c** could also be written as $\frac{x - 2}{3} = 15$ and solved in exactly the same way as in the example.

Notice that multiplying out the brackets of $\frac{1}{3}(x - 2) = 15$ first would give $\frac{1}{3}x - \frac{2}{3} = 15$. Then you would have to deal with coefficients that are fractions. The method shown above is usually easier.

Exercise 15A

1 Copy and complete these equations, given that $x = 7$.

   a $4x + 5 = \ldots$ **b** $4(x + 5) = \ldots$ **c** $\frac{1}{2}(x + 5) = \ldots$ **d** $\frac{x + 5}{3} = \ldots$

2 Solve these equations. If the solution is not a whole number, write it as a mixed number.

   a $2x - 12 = 14$ **b** $5y + 4 = 49$ **c** $4z - 16 = 7$ **d** $3t + 6 = 23$

   e $7x - 3 = 50$ **f** $3p + 14 = 24$ **g** $5k - 11 = 2$ **h** $8n + 3 = 40$

3 Solve these equations.

   a $2(x - 3) = 16$ **b** $4(x + 5) = 32$ **c** $3(y - 6) = 36$ **d** $8(t + 2) = 72$

   e $7(5 + t) = 140$ **f** $12(p - 4) = 36$ **g** $50 = 2(x + 8)$ **h** $3(a - 4) = 120$

4 Solve these equations.

   a $\frac{1}{4}x = 12$ **b** $\frac{1}{4}x + 3 = 12$ **c** $\frac{1}{4}(x + 3) = 12$ **d** $\frac{x + 4}{3} = 12$

5 Solve these equations.

   a $\frac{1}{3}y - 5 = 7$ **b** $\frac{1}{2}t + 4 = 13$ **c** $\frac{y - 8}{2} = 11$ **d** $\frac{1}{4}(t + 2) = 15$

   e $\frac{1}{3}(6 + x) = 4$ **f** $\frac{t + 11}{6} = 4$ **g** $8 = \frac{1}{2}(t - 11)$ **h** $\frac{1}{8}(x + 10) = 4$

6 Here is an equation:

   $3(x - 5) = 11$

   a Solve the equation by multiplying out the brackets first.

   b Solve the equation by dividing by 3 first.

7 Solve these equations. Give the answers as mixed numbers.

   a $4x - 12 = 15$ **b** $4(x - 3) = 9$ **c** $4(z + 3) = 14$ **d** $8r + 7 = 19$

   e $12x - 15 = 12$ **f** $6(x + 3\frac{1}{2}) = 39$ **g** $2(y - 1\frac{1}{4}) = 11$ **h** $10r + 5 = 69$

8 The perimeter of this trapezium is $37\frac{1}{2}$ cm.

   a Write down an equation to show this.

   b Solve the equation to find the value of x.

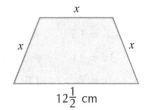

$12\frac{1}{2}$ cm

9 Solve these equations. Write the answers as decimals.

   a $2x + 4.1 = 11.3$ **b** $5y - 18.6 = 16.9$ **c** $3(z - 8.2) = 12.6$ **d** $1.4(w + 6.2) = 12.6$

   e $\frac{t - 1.7}{3.2} = 4.5$ **f** $0.32(x - 1.3) = 0.8$ **g** $55d + 39 = 61$ **h** $1.8t + 32 = 144.5$

Challenge: Odd one out

Four of these equations have the same solution.

$5(x - 4) = 75$ $5x - 30 = 75$ $6(x - 6.5) = 75$ $18 + 3x = 75$ $3(6 + x) = 75$

Which is the odd one out? Justify your answer.

15.2 Equations with the variable on both sides

Learning objective

- To solve equations with the variable on both sides

Jake is thinking of a number.

Call Jake's number x.

Then you can write this equation:

$x + 12 = 3x - 45$

This equation has x on both sides. Example 2 shows how to solve this equation.

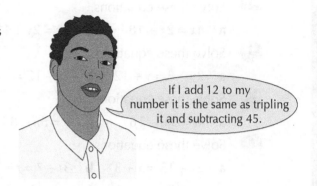

If I add 12 to my number it is the same as tripling it and subtracting 45.

Example 2

Solve the equation $x + 12 = 3x - 45$.

$x + 12 = 3x - 45$	First, subtract x from both sides.
$12 = 2x - 45$	This removes the x from the left-hand side. On the right, $3x - x = 2x$.
	Now x only appears on one side. Solve this in the usual way.
$57 = 2x$	Add 45 to both sides. $12 + 45 = 57$
$28.5 = x$	Divide by 2. Half of 57 is 28.5.

When the variable occurs on both sides of the equal sign, remove it from one side by adding or subtracting.

Example 3

Solve these equations.   **a** $2x - 4 = 5x - 24$ **b** $4y + 8 = 43 - y$

a $2x - 4 = 5x - 24$	Subtract $2x$ from both sides.
$-4 = 3x - 24$	You now have -4 on the left-hand side. Add 24 to both sides.
$20 = 3x$	$-4 + 24 = 20$. Now divide by 3.
$x = \frac{20}{3} = 6\frac{2}{3}$	It is better to leave the answer as a fraction in this case.
b $4y + 8 = 43 - y$	To remove the $-y$ from the right-hand side, add y to both sides.
$5y + 8 = 43$	$4y + y = 5y$. Solve this new equation in the usual way.
$5y = 35$	Subtract 8 from both sides. $43 - 8 = 35$
$y = 7$	Divide by 5 to find the value of y.

You can check this is correct.

When $y = 7$, then $4y + 8 = 4 \times 7 + 8 = 36$ and $43 - y = 43 - 7 = 36$.

They have the same value.

1 Solve these equations.

a $2x = x + 15$ b $4x = x + 45$ c $3t = t + 24$ d $5x = x + 44$

2 Solve these equations.

a $4x = 2x + 18$ b $5y = 2y + 18$ c $6t = 2t + 18$ d $6k = 5k + 18$

3 Solve these equations.

a $2y - 8 = y + 12$ b $4z - 17 = z + 4$ c $3x - 14 = x + 8$ d $6x - 13 = 2x + 9$

4 Solve these equations.

a $m = 3m - 16$ b $n = 5n - 84$ c $2p = 5p - 18$ d $4x = 10x - 48$

5 Solve these equations.

a $2x - 15 = x + 3$ b $3t - 7 = t + 14$ c $5x - 4 = 2x + 6$ d $n + 15 = 2n - 19$

e $y + 12 = 2y - 14$ f $2x + 29 = 6x + 26$ g $5d - 13 = d - 5$ h $2x - 13 = 3x - 21$

i $3 + 4x = 17 + 2x$ j $5d - 27 = 3d - 1$ k $x = 5x - 20$ l $2.5x + 6 = 3x - 4$

PS **6** The perimeters of these shapes are the same length.

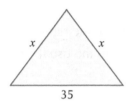 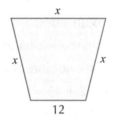

a Write an equation to show this.

b Solve the equation.

c Work out the perimeter of each shape.

PS **7** Ann has £n. Carrie has £75 more than Ann.

a Write down an expression, in terms of n, for the amount of money Carrie has.

b Carrie has four times as much as Ann. Write down an equation to show this.

c Solve the equation and work out how much each person has.

8 Given that $x = 5$, work out the value of each expression.

a $4x - 6$ b $24 - 3x$ c $30 - 4x$ d $8 - \frac{1}{2}x$

9 Solve these equations.

a $x = 24 - x$ b $2x = 24 - x$ c $3x = 24 - x$ d $5x = 24 - x$

10 Solve these equations.

a $x - 12 = 18 - x$ b $2x - 5 = 22 - x$ c $4x + 3 = 27 - 2x$ d $40 - 2x = 4 + x$

e $3x - 1 = 19 - 2x$ f $x + 28 = 100 - 2x$ g $6x - 4 = 10 - x$ h $31 - 2\frac{1}{2}x = 16 + \frac{1}{2}x$

11 Solve these equations.

a $x + 11 = 3x + 2$ b $x + 2 = 11 - 3x$ c $x + 2 = 3x - 11$ d $11 - x = 3x - 2$

Challenge: Muddying the waters

Look at this equation. $\qquad\qquad$ $3x + 4 = 17$

If you add 1 to both sides you get: $\qquad$ $3x + 5 = 18$

If you add 2 to both sides you get: $\qquad$ $3x + 6 = 19$

A Copy and complete these equations in the same way.

$3x + 8 = \ldots$

$3x + 13 = \ldots$

$3x + \ldots = 30$

$3x + \ldots = 45$

Here is the original equation again. $\qquad$ $3x + 4 = 17$

If you add x to both sides you get: $\qquad$ $4x + 4 = 17 + x$

Add another x: $\qquad\qquad\qquad\qquad$ $5x + 4 = 17 + 2x$

B Copy and complete these equations in a similar way.

$8x + 4 = \ldots$

$12x + 4 = \ldots$

$17x + 4 = \ldots$

$\ldots = 17 + 12x$

15.3 More complex equations

Learning objectives

- To solve equations with fractional coefficients
- To solve equations with brackets and fractions

If you have brackets in an equation where the variable occurs more than once, it is usually best to multiply out the brackets first.

Example 4

Solve the equation $3(t - 11) = 2(t + 8)$.

$3(t - 11) = 2(t + 8)$ $\qquad$ There are two sets of brackets. Multiply out both of them.

$3t - 33 = 2t + 16$ $\qquad$ Now subtract $2t$ from both sides.

$t - 33 = 16$ $\qquad\qquad$ By subtracting $2t$ you remove the t-term from one side.

$t = 49$ $\qquad\qquad\quad$ Add 33 to both sides. $16 + 33 = 49$

You can check this is correct.

$3(49 - 11) = 3 \times 38 = 114$ and $2(49 + 8) = 2 \times 57 = 114$ so they are the same.

If there is a fraction in front of a bracketed term, you can get rid of it by multiplying by the denominator.

Example 5

Solve the equation $\frac{2}{3}(x + 8) = x$.

$\frac{2}{3}(x + 8) = x$ First, multiply by 3.

$2(x + 8) = 3x$ Now multiply out the brackets.

$2x + 16 = 3x$ Now subtract $2x$ from both sides.

$16 = x$ $3x - 2x = x$

Exercise 15C

1   Solve these equations.

    a $2(x - 4) = 10$ **b** $2(x - 4) = x$ **c** $2(x - 4) = x + 10$ **d** $2(x + 4) = x + 10$

2   Solve these equations.

    a $3(y - 5) = 2y + 9$ **b** $a + 15 = 3(a - 1)$ **c** $2(t + 12) = 5t$

    d $18 + x = 3(x - 8)$ **e** $2(n + 6) = 3n - 10$ **f** $3x = 2(20 - x)$

    g $2(x - 3) = 15 - x$ **h** $4(12 - x) = 8 + x$

3   **a**   Copy and complete this table.

x	5	6	7	8	9
$2(x + 1)$	12				
$3(x - 2)$	9				

    b   Use the table to solve the equation $2(x + 1) = 3(x - 2)$.

    c   Solve the equation algebraically and check that you get the same answer.

4   **a**   Copy and complete this table.

x	1	2	3	4	5
$4(x + 2)$	12				
$3(12 - x)$	33				

    b   Use the table to solve the equation $4(x + 2) = 3(12 - x)$.

    c   Solve the equation algebraically and check that you get the same answer.

5   Solve these equations.

    a $3(x - 6) = 2(x + 3)$ **b** $2(a + 9) = 3(a - 1)$ **c** $4(t - 3) = 2(t + 8)$

    d $5(1 + x) = 3(x + 7)$ **e** $2(y - 1) = 3(y - 6)$ **f** $3(k - 4) = 2(9 - k)$

    g $4(10 - p) = 2(5 + p)$ **h** $10(x + 4) = 2(x + 20)$ **i** $5(x - 4) = 3(x + 7)$

6 Here are two straight lines.

The first line is divided into five equal parts, each is $x - 4$ cm long.

The second line is divided into three equal parts, each is $x + 2$ cm long

a The lines are the same length. Write an equation to show this.

b Solve the equation.

c Work out the length of each line.

 7 **a** Write down an expression for the length of the perimeter of the triangle.

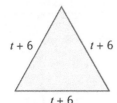 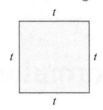

b Write down an expression for the length of the perimeter of the square.

c The perimeters of the shapes are the same length. Write down an equation to show this.

d Solve the equation.

e Work out the lengths of the sides of each shape.

 8 **a** Write down an expression for the area of the first rectangle.

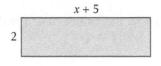 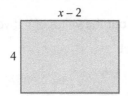

b Write down an expression for the area of the second rectangle.

c The shapes have the same area. Write down an equation to show this.

d Solve the equation.

e Work out the area of each shape.

9 Solve these equations.

a $\dfrac{x - 5}{3} = 2$
    b $\dfrac{y + 8}{4} = 5$
    c $\dfrac{t + 8}{2} = 6$

d $\dfrac{4 + x}{3} = 4$
    e $\dfrac{y + 12}{5} = 5$
    f $\dfrac{x - 6}{4} = 3$

g $\dfrac{x + 1}{2} = 12$
    h $\dfrac{n - 4}{4} = 7$
    i $\dfrac{x - 12}{8} = 2$

 10 Alex is a years old.

a Write down an expression for Alex's age in six years' time.

b In six years' time, Alex's age divided by 4 will be 12.

Write down an equation to show this.

c Solve the equation. How old is Alex?

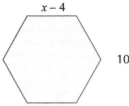
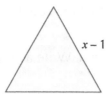
15.4 Rearranging formulae

Learning objective

* To change the subject of a formula

This is a formula that is used in science.

$v = u + at$

The **subject** of the formula is v.

The formula shows you how to work out the value of v if you know the values of u, a and t.

Subtracting at from both sides, you get:

$v - at = u$

Now write this the other way round:

$u = v - at$

Now u is the subject of the formula.

Example 6

Rearrange the formula $v = u + at$ to make a the subject.

$v = u + at$	You need to get a on its own, on one side of the equals sign.
$v - u = at$	First, subtract u from both sides.
$\dfrac{v - u}{t} = a$	Now divide by t. Remember that at means $a \times t$.
The required formula is $a = \dfrac{v - u}{t}$.	Write it the other way round, so that a is on the left.

Exercise 15D

1 This is a formula you will see in science.

$$v = u + 10t$$

  a Work out the value of v, given that $u = 40$ and $t = 3$.

  b Work out the value of v, given that $u = 15$ and $t = 2.5$.

  c When $v = 60$ and $t = 2$, what is the value of u?

  d When $v = 100$ and $t = 6$, what is the value of u?

2 Make t the subject of each formula.

  a $s = t + 25$ **b** $w = t - 6.5$ **c** $a = 4t$ **d** $m = \frac{1}{2}t$

3 Make n the subject of each formula.

  a $T = m + n$ **b** $q = n + t - 12$ **c** $y = 3n + a$ **d** $y = 3(n + 1)$

4 This is the formula for the mean, m, of two numbers, x and y.

$$m = \frac{x + y}{2}$$

  a Work out the value of m, given that $x = 3.5$ and $y = 12.5$.

  b Work out the value of m, given that $x = 124$ and $y = 138$.

  c Show that the formula can be rearranged as $x = 2m - y$.

  d Use the formula in part **c** to find the value of x, given that $m = 41$ and $y = 48$.

  e Rearrange the formula to make y the subject.

  f Use your formula from part **e** to find the value of y, given that $m = 8.5$ and $x = 6.3$.

5 This is the formula from question **1** again.

$$v = u + 10t$$

  a Rearrange the formula to make u the subject.

  b Use your answer to part **a** to find the value of u when $v = 19.4$ and $t = 0.8$.

  c Rearrange the formula to make t the subject.

  d Use your answer to part **c** to find the value of t when $u = 67$ and $v = 91$.

6 This is a formula for the area, a, of this shape.

$$a = 4x + 30$$

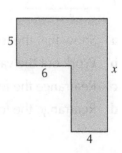

  a Work out the value of a when:

     i $x = 7$ **ii** $x = 20$.

  b Rearrange the formula to make x the subject.

  c Work out the value of x, given that:

     i $a = 50$ **ii** $a = 70$.

7   The equation of a straight line is a formula connecting x and y.

Make x the subject of each of these equations of straight lines.

   a $y = 5x + 12$ b $y = \frac{1}{3}x - 2$ c $x + y = 50$

   d $y = 45 + 25x$ e $x + 3y = 20$ f $4x + y = 18$

8   Look at this formula.

   $$k = a + 3b - 1$$

   a   Rearrange the formula to make a the subject.

   b   Rearrange the formula to make b the subject.

9   This is a formula from Example 6.

   $$v = u + at$$

   a   Work out the value of v when $u = 10$, $a = 3$ and $t = 5$.

   b   Rearrange the formula to make t the subject.

   c   Use your formula in part **b** to work out the value of t when $v = 20$, $u = 8$ and $a = 2$.

10 a   Rearrange the perimeter formula to make a the subject.

   b   Rearrange the perimeter formula to make b the subject.

   c   Rearrange the area formula to make a the subject.

   d   Rearrange the area formula to make b the subject.

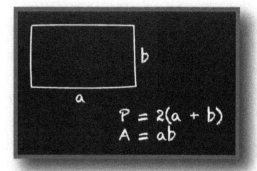

11 This is an isosceles triangle.

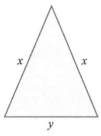

   a   Show that the formula for the perimeter, p, is $p = 2x + y$.

   b   Work out the value of p when $x = 14$ and $y = 10$.

   c   Rearrange the formula to make y the subject.

   d   Rearrange the formula to make x the subject.

12 Here is a formula.

$$t = 10p + 20q$$

   a Rearrange the formula to make p the subject.

   b Rearrange the formula to make q the subject.

Financial skills: Paying interest

If you take out a loan you must pay interest to the lender.

If the amount of the loan is £A and the rate of interest is R%, the amount of interest you must pay, £I, is given by the formula:

$$I = \frac{AR}{100}$$

You must pay this interest, as well as paying back the loan.

A **a** Work out the interest you must pay on a loan of £400 at 18% interest.

    b Work out the total amount you must pay back.

B **a** Work out the interest you must pay on a loan of £2000 at 3% interest.

    b Work out the total amount you must pay back.

C **a** Make R the subject of the formula $I = \frac{AR}{100}$.

    b Use this formula to work out the rate of interest for someone who has a loan of £400 and has to pay £64 interest.

Ready to progress?

I can solve equations involving brackets and fractions.
I can solve equations where the variable occurs on both sides.

I can change the subject of a formula in simple cases.

Review questions

1 Solve these equations.

   a $2(x + 3) = 15$ b $\frac{1}{2}(y - 5) = 8$ c $\frac{1}{3}t - 9 = 6$ d $25 = 11 + 4x$

2 Solve these equations.

   a $3x - 5 = x + 11$ b $5x - 23 = 4x - 6$

   c $3 + 2y = 6y - 21$ d $w + 12 = 21 - w$

3 Solve these equations.

   a $2(x - 15) = x - 8$ b $5(x - 6) = 10 + x$

   c $3(t - 7) = t + 19$ d $2(m + 8) = 3(m - 6)$

4 Solve these equations.

   a $\dfrac{x + 8}{2} = x - 5$ b $\frac{2}{3}(y + 4) = y - 4$

   c $\frac{3}{4}(t + 5) = t - 2$ d $\frac{2}{5}(t - 3) = t - 6$

 5 a   The sum of the angles of any triangle is 180°. Write an equation in terms of x.

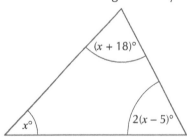

   b   Work out the angles of the triangle.

6 **a** Write down expressions for the areas of rectangle A and rectangle B.

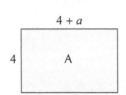
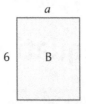

b The rectangles have the same area. Write down an equation to show this.

c Solve the equation and work out the area of each rectangle.

d Do the rectangles have the same perimeter? Justify your answer.

7 The formula for the area, A, of this triangle is:

$A = \dfrac{bh}{2}$

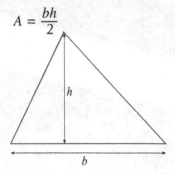

a Rearrange the formula to make b the subject.

b Rearrange the formula to make h the subject.

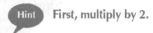

Hint First, multiply by 2.

8 Look at this formula.

$f = 1.8c + 32$

Rearrange the formula to make c the subject.

9 This is the equation of a line.

$y = \frac{1}{2}x + 5$

Rearrange it to make x the subject.

10 Look at this formula.

$p = 2a + 4b$

a Rearrange it to make a the subject.

b Rearrange it to make b the subject.

11 This is the formula for the mean, m, of a and b.

$m = \dfrac{a + b}{2}$

a Rearrange the formula to make a the subject.

b Rearrange the formula to make b the subject.

Reasoning
Using graphs to solve equations

You can use graphs to solve equations.

1 a Solve the equation $2(x + 1) = 8 - x$.

b Copy and complete this table of values.

x	0	1	2	3	4	5
$2(x + 1)$				8		
$8 - x$				5		

c Draw a pair of axes like this.

d Use the table of values to draw graphs of:

   i $y = 2(x + 1)$

   ii $y = 8 - x$.

e Find the x-coordinate of the point where the lines cross. Check that this is the solution of the equation $2(x + 1) = 8 - x$, from your answer to part **a**.

2 Here are some equations. In each case:

  i solve the equations by finding the
    x-coordinate of the point where two
    appropriate lines on the graph cross

  ii check your answer by solving the
    equations algebraically.

  a $4x = x + 3$ **b** $2(x - 4) = x$

  c $4x = 12 - 2x$ **d** $2(x - 4) = 7$

16

Comparing data

This chapter is going to show you:

- how to construct grouped frequency tables for data
- how to interpret and draw grouped frequency diagrams
- how to compare two distributions by using an average and the range
- how to select the correct average when analysing data.

You should already know:

- how to construct a frequency table from given data
- how to find the mode, median, range and modal class for grouped data
- how to calculate the mean from a set of data
- how to construct graphs and diagrams to represent data.

About this chapter

How can a coach select the best members for any sporting team, whether for the Olympics or a local football club? It helps to know and compare the records of all the applicants – not only their best times, points or scores over the previous few months but also how consistent their performances are.

You can use the same techniques to help you find the place that has the best record for high temperatures and sunshine for a good beach holiday. This chapter shows you how you can use data in this way to help make decisions.

16.1 Grouped frequency tables

Learning objective

• To create a grouped frequency table from raw data

Key words

class	continuous
discrete	frequency
grouped frequency table	

You can draw up a frequency table to record how many times each value in a set of data occurs. The number of times a data value occurs is its **frequency**.

If the data is about specific events it is easy to see a pattern, for example, when people vote for different political candidates in an election there is always a limited selection of choices to be made. This is called **discrete** data.

Often, data has a wide range of possible values; examples include masses or heights of all the pupils in a class. This is called **continuous** data and you have to group it together, to see any pattern. In a **grouped frequency table**, you arrange information into **classes** or groups of data to do this. You can create frequency diagrams, from grouped frequency tables, to illustrate the data.

Example 1

These are the journey times, in minutes, for a group of 16 railway travellers.

| 25 | 47 | 12 | 32 | 28 | 17 | 20 | 43 | 15 | 34 | 45 | 22 | 19 |
| 36 | 44 | 17 |

Construct a frequency table to represent the data.

Looking at the data, 10 minutes is a sensible class interval size.

> Write the class intervals in the form $10 < T \leqslant 20$.
>
> $10 < T \leqslant 20$ is a short way of writing the time interval of 10 minutes to 20 minutes. The possible values for T include 20 minutes but not 10 minutes.

There are six times in the group $10 < T \leqslant 20$.   12, 17, 20, 15, 19 and 17

There are three times in the group $20 < T \leqslant 30$.   25, 28 and 22

There are three times in the group $30 < T \leqslant 40$.   32, 34 and 36

There are four times in the group $40 < T \leqslant 50$.   47, 43, 45 and 44

Put all this information into a table.

Time, T (minutes)	Frequency
$10 < T \leqslant 20$	6
$20 < T \leqslant 30$	3
$30 < T \leqslant 40$	3
$40 < T \leqslant 50$	4

1 This table shows the lengths of time 25 customers spent in a shop.

a One of the customers was in the shop for exactly 20 minutes. In which class was this customer recorded?

b Which is the modal class?

Time, T (minutes)	Frequency
$0 < T \leqslant 10$	12
$10 < T \leqslant 20$	7
$20 < T \leqslant 30$	6

2 These are the heights (in metres) of 20 people.

1.65	1.53	1.71	1.72
1.48	1.74	1.56	1.55
1.80	1.85	1.58	1.61
1.82	1.67	1.47	1.76
1.79	1.66	1.68	1.73

a Copy and complete the frequency table.

b What is the modal class?

Height, h (metres)	Frequency
$1.40 < h \leqslant 1.50$	
$1.50 < h \leqslant 1.60$	
$1.60 < h \leqslant 1.70$	
$1.70 < h \leqslant 1.80$	
$1.80 < h \leqslant 1.90$	

3 These are the masses (in kilograms) of fish caught in one day by an angler.

0.3	5.6	3.2	0.4
0.6	1.1	2.4	4.8
0.5	1.6	5.1	4.3
3.7	3.5		

a Copy and complete the frequency table.

b What is the modal class?

c The angler's daughter arrived with some lunch just as he was catching a fish. What is the probability that this fish was in the range of $3 < M \leqslant 4$?

Mass, M (kilograms)	Frequency
$0 < M \leqslant 1$	
$1 < M \leqslant 2$	
...	
...	
...	
...	

4 These are the temperatures (in Celsius degrees, °C) of 16 towns in Britain on one day.

12	10	9	13	12	14	17
16	18	10	12	11	15	15
12	13					

a Copy and complete the frequency table.

b What is the modal class?

c What is the range of temperatures in these cities?

Temperature, T (°C)	Frequency
$8 < T \leqslant 10$	
$10 < T \leqslant 12$	
...	
...	
...	

PS **5** A petrol station owner surveyed a sample of customers to see how many litres of petrol they bought. These are the results.

27.6	31.5	48.7	35.6	44.8	56.7	51.0	39.5
28.8	43.8	47.3	36.6	42.7	45.6	32.4	51.7
55.9	44.6	36.8	49.7	37.4	41.2	38.5	45.9
34.1	54.3	41.3	49.4	38.7	33.2		

Using a class size of 5 litres, work out the modal class of the amount of petrol that customers bought.

PS **6** In a doctors' surgery, the practice manager told each doctor that the length of most consultations should be more than 5 minutes but less than 10. She monitored the consultation times of the three doctors at the practice throughout one day. These are the results.

Dr Speed (minutes): 6, 8, 11, 5, 8, 5, 8, 10, 12, 4, 3, 6, 8, 4, 3, 15, 9, 2, 3, 5

Dr Bell (minutes): 7, 12, 10, 9, 6, 13, 6, 7, 6, 9, 10, 12, 11, 14

Dr Khan (minutes): 5, 9, 6, 3, 8, 7, 3, 4, 5, 7, 3, 4, 5, 9, 10, 3, 4, 5, 4, 3, 4, 4, 9

a Did any of the doctors manage to follow the practice manager's advice?

b Write a short report about the consultation times of the three doctors.

Activity: Textbook text

A Collect a large number of school textbooks.

B Record the number of pages in each book.

C Choose suitable class intervals for the data.

D Record the data in a frequency table.

E Comment on your results.

16.2 Drawing frequency diagrams

Learning objectives

- To interpret frequency diagrams

- To draw a frequency diagram from a grouped frequency table

Key words	
block graph	frequency diagram
line graph	

Look at the picture. How could the organisers record the finishing times to show when most of the runners finish?

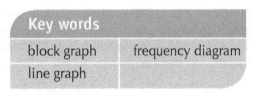

They could draw a frequency diagram such as a block graph or a line graph as seen in the following examples.

Example 2

Construct a frequency diagram for this data about runners' race times.

Race times, t (minutes)	Frequency
$0 < t \leqslant 15$	4
$15 < t \leqslant 30$	5
$30 < t \leqslant 45$	10
$45 < t \leqslant 60$	6

It is important that the diagram has a title and labels, as shown here.

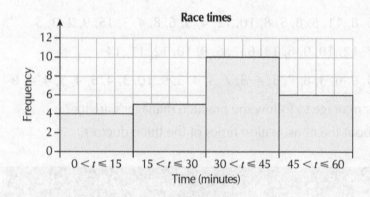

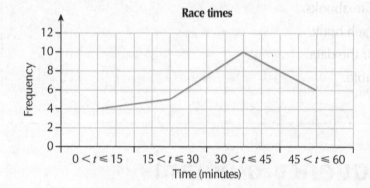

Note that you can illustrate the data in several different ways. A block graph and a line graph have both been used in the previous example. The line graph uses the heights of the block graph to show the variation of the data.

Example 3

A shop manager collected data about the daily sales of ice cream every month throughout the year. He worked out the average (mean) daily sales figure for each month and put it on the graph below. In which month were average sales at their highest? Give a reason why you think this happened.

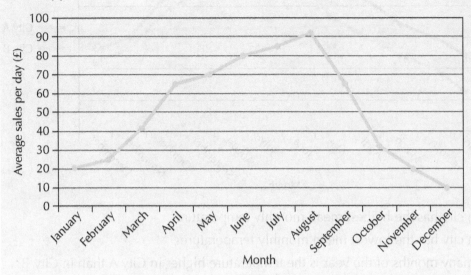

The highest average sales were in August (£92 per day). This was probably because the weather was warmer, as people tend to buy ice cream in warm weather.

Exercise 16B

1 Construct a frequency diagram from each frequency table.

a Aircraft flight times

Time, T (hours)	Frequency
$0 < T \leqslant 1$	3
$1 < T \leqslant 2$	6
$2 < T \leqslant 3$	8
$3 < T \leqslant 4$	7
$4 < T \leqslant 5$	4

b Temperatures of European capital cities

Temperature, T (°C)	Frequency
$0 < T \leqslant 5$	2
$5 < T \leqslant 10$	6
$10 < T \leqslant 15$	11
$15 < T \leqslant 20$	12
$20 < T \leqslant 25$	7

c Lengths of metal rods

Length, l (centimetres)	Frequency
$0 < l \leqslant 10$	9
$10 < l \leqslant 20$	12
$20 < l \leqslant 30$	6
$30 < l \leqslant 40$	3

d Masses of animals on a farm

Mass, M (kg)	Frequency
$0 < M \leqslant 20$	15
$20 < M \leqslant 40$	23
$40 < M \leqslant 60$	32
$60 < M \leqslant 80$	12
$80 < M \leqslant 100$	6

2 This graph shows the mean monthly temperature for two cities.

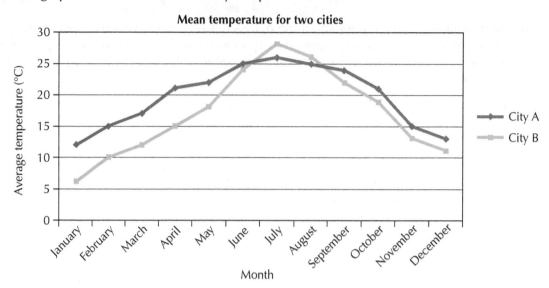

a Which city has the highest mean monthly temperature?

b Which city has the lowest mean monthly temperature?

c How many months of the year is the temperature higher in City A than in City B?

d What is the difference in average temperature between the two cities in February?

PS **3** The three classes of a year group collected sponsor money for charity. These are the amounts collected by individual pupils.

Class A (£): 3, 9.60, 5.50, 8.45, 8, 7.35, 1.55, 15.50, 19, 14, 12.75, 13.50, 11.85, 12, 14.36, 15.40, 17, 6.55, 7.40, 8, 6.32, 1, 3.65, 16.50, 14, 19.55, 18, 16.46, 19

Class B (£): 5, 11.50, 7.40, 10.55, 10, 9.15, 3.55, 18.70, 21, 116, 14.95, 15.70, 13.95, 14, 16.30, 17.50, 19, 8.75, 9.20, 10, 8.12, 3, 5.85, 18.60, 16, 21.95, 20, 18.26, 21, 14.35, 5.47, 13.55

Class C (£): 4, 8.50, 6.70, 7.35, 9, 6.25, 2.75, 14.30, 20, 13, 11.65, 14.70, 10.45, 13, 13.38, 14.30, 18, 5.45, 8.70, 7, 7.42, 2, 2.15, 17.30, 13, 20.65, 17, 17.56, 18, 17.55, 1965, 17.36

Create a diagram that the head of year can use, to show how well the pupils had done in raising sponsorship money, so that he can encourage the school to collect for charity.

Activity: Comparing holiday destinations

Use travel brochures or the internet to compare the temperatures at two popular European destinations.

Design a poster to advertise one destination as being more appropriate than the other, if you want a warm holiday.

16.3 Comparing data

Learning objective

- To use mean and range to compare data from two sources

The picture shows several shots made by a golfer. What is the range of the shots?

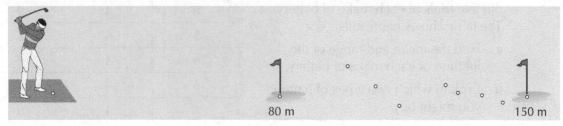

It is often important to know the range of results as this can show how consistent they are. A golfer whose shots ranged over 20 m to 150 m would be performing less consistently than this one.

Example 4

The table shows the mean and range of two teams' basketball scores.

Compare the mean and range and explain what they tell you.

	Team A	Team B
Mean	75	84
Range	20	10

The means tell you that the average score for Team B is higher than that for Team A, so they have higher scores generally.

The range compares the differences between the lowest and highest scores. As this is higher for Team A, there is more variation in their scores. You could say that they are less consistent.

Exercise 16C

1. A factory worker recorded the start and finish times of a series of jobs.

Job number	1	2	3	4	5
Start time	9:00 am	9:20 am	9:50 am	10:10 am	10:20 am
Finish time	9:15 am	9:45 am	10:06 am	10:18 am	10:38 am

   a Work out the range of the times taken for the jobs.

   b Calculate the mean length of time taken for the five jobs.

2. These are the minimum and maximum temperatures for four English counties, in April.

County	Northumberland	Leicestershire	Oxfordshire	Surrey
Minimum (°C)	2	4	4	4.5
Maximum (°C)	12	15	16.5	17.5

   a Find the range of the temperatures for each county.

   b Comment on any differences you notice, explaining why these differences might occur.

3 The table shows the mean and range of a set of test scores for Jon and Matt.

Compare the mean and range and explain what they tell you.

	Jon	Matt
Mean	64	71
Range	35	23

MR **4** Fiona recorded how long, to the nearest hour, Everlast, Powercell and Electro batteries lasted in her MP3 player. She did five trials of each make of battery. The table shows her results.

Everlast (£1.00 each)	Powercell (50p each)	Electro (£1.50 each)
6	4	9
5	6	8
6	3	9
6	4	9
7	4	9

a Find the mean and range of the lifetime of each make of battery.

b Explain which two types of battery you might buy.

5 This table shows the average daily maximum temperatures for Cardiff, Edinburgh and London over one year.

	Jan	Feb	Mar	Apr	May	Jun	Jul	Aug	Sep	Oct	Nov	Dec
Cardiff (°C)	7	7	10	13	16	19	20	21	18	14	10	8
Edinburgh (°C)	6	6	8	11	14	17	18	18	16	12	9	7
London (°C)	6	7	10	13	17	20	22	21	19	14	10	7

Using the means and ranges of these maximum temperatures, write a report comparing the temperatures in the three cities.

Activity: Comparing populations

Use the internet to find the populations of the four largest cities in China and the four largest cities in the United States of America.

Use the mean and the range to compare the populations in these cities.

16.4 Which average to use?

Learning objective

- To understand when each different type of average is most useful

How would you work out an average height for this group of people?

You cannot use the mode because all their heights are different, so none is more common than the rest.

You can use the median – note that this doesn't take into account how small the smallest value is.

The best type of average to use is the mean, which takes all the heights into account. You find this by adding all the values and dividing by the number of values.

There is a big difference between the shortest and tallest people. Therefore you could also use the range of heights to give some idea of how close to the average each person might be.

This table will help you decide which type of average to use for a set of data.

	Advantages	Disadvantages	Example
Mean	Uses every piece of data. Probably the most used average.	May not be representative when the data contains an extreme value.	1, 1, 1, 2, 4, 15 $\text{Mean} = \dfrac{1+1+1+2+4+15}{6} = 4$ which is a higher value than most of the data.
Median	Only looks at the middle values, so it is a better average to use if the data contains extreme values.	Not all values are considered so could be misleading.	1, 1, 3, 5, 10, 15, 20 Median = 4th value = 5 Note that above the median the numbers are a long way from the median but below the median they are very close.
Mode	It is the most common value.	If the mode is an extreme value it is misleading to use it as an average.	Weekly wages of a boss and his four staff: £150, £150, £150, £150, £1000. Mode is £150 but mean is £320.
Modal class for continuous data	This is the class with the greatest frequency.	The actual values may not be centrally placed in the class.	<table><tr><th>Time (T) minutes</th><th>Frequency</th></tr><tr><td>$0 < T \leqslant 5$</td><td>2</td></tr><tr><td>$5 < T \leqslant 10$</td><td>3</td></tr><tr><td>$10 < T \leqslant 15$</td><td>6</td></tr><tr><td>$15 < T \leqslant 20$</td><td>1</td></tr></table> The modal class is $10 < T \leqslant 15$, but all six values may be close to 15.
Range	It measures how spread out the data is.	It only looks at the two extreme values, which may not represent the spread of the rest of the data.	1, 2, 5, 7, 9, 40 The range is 40 − 1 = 39 but without the last value (40), the range would be only 8.

1 **i** Calculate the given average for each set of data.

  ii Explain whether the given average is a sensible one to use for that set of data.

   a The mean of 2, 3, 5, 7, 8, 10   **b** The mode of 0, 1, 2, 2, 2, 4, 6

   c The median of 1, 4, 7, 8, 10, 11, 12   **d** The mode of 2, 3, 6, 7, 10, 10, 10

   e The median of 2, 2, 2, 2, 4, 6, 8   **f** The mean of 1, 2, 4, 6, 9, 30

2 These are the times (in seconds) that 15 pupils took to complete a short task.

10.1, 11.2, 11.5, 12.1, 12.3, 12.8, 13.6, 14.4, 14.5, 14.7, 14.9, 15.4, 15.9, 16.6, 17.1

  a Copy and complete the frequency table and find the modal class.

Time, T (seconds)	Tally	Frequency
$10 < T \leqslant 12$		
$12 < T \leqslant 14$		
$14 < T \leqslant 16$		
$16 < T \leqslant 18$		

  b Explain why the mode is unsuitable for the ungrouped data, but the modal class is suitable for the grouped data.

3 **i** Calculate the range for each set of data.

  ii Decide whether it is a suitable measure of the spread. Explain your answer.

   a 1, 2, 4, 7, 9, 10   **b** 1, 10, 10, 10, 10   **c** 1, 1, 1, 2, 10

   d 1, 3, 5, 6, 7, 10   **e** 1, 1, 1, 7, 10, 10, 10   **f** 2, 5, 8, 10, 14

(MR) **4** A group returning from a trip abroad were asked to donate their spare change to charity. This is what they were able to donate.

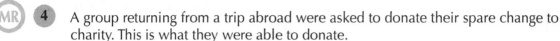

$0.80 $3.50 $1.60 $5.22 $0.42 $0.06 $3.15 $4.38 $2.72

$0.70 $2.90 $5.45 $1.32 $2.05 $0.67 $5.21 $4.57 $2.30

$4.18 $0.19 $0.01 $5.86 $3.17 $5.08 $3.76 $3.14 $2.19

  a Copy and complete the frequency table for these figures.

Donation, M ($)	Tally	Frequency
$0 < M \leqslant 1$		
$1 < M \leqslant 2$		
$2 < M \leqslant 3$		
$3 < M \leqslant 4$		
$4 < M \leqslant 5$		
$5 < M \leqslant 6$		

b What is the modal class?

c Find the median amount donated.

d Explain why the modal amount donated is a better average to use than the median.

 5 A factory employs 500 people. Of these, 490 are workers who each earn less than £250 a week and 10 are managers earning more than £900 a week each.

a Which average would you use to argue that pay at the factory was low?

b Which average would you use to argue that pay at the factory was reasonable?

c In discussions about average pay, which average would be used by:

   i the worker   **ii** the manager   **iii** the owners of the factory?

Give reasons for your answers.

 6 In an ice-skating competition, the judges gave these scores.

Craig	4	6	2	7	6	5	7	6	4	6
Len	6	7	6	8	7	7	8	7	7	8
Darcy	6	6	5	8	7	6	8	7	6	7
Bruno	7	7	6	8	8	7	9	7	7	8

Use the mean score, as well as the range, to compare the scoring of each judge.

  7 A snooker team needs a replacement player for an important match. The captain decides to look at the last nine scores of two possible replacements. These are their scores.

Joe	48	79	53	88	75	64	72	49	65
Jimmy	110	30	36	119	25	31	28	101	41

a Imagine you were Joe. Give your reasons for being chosen.

b Imagine you were Jimmy. Give your reasons for being chosen.

c If you were the captain of the snooker team, whom would you choose? Why?

Activity: At the London Arena

A Collect a set of data about the attendances at the London Arena for different singers and groups. Calculate the mean, median and mode.

B Repeat this exercise for a large arena close to where you live.

C Compare the differences in the distributions of the data. Explain any differences you find in the averages.

Ready to progress?

I can compare the range and the mean from two sets of data.
I can decide which average is the best to use in different circumstances.

I can construct grouped frequency tables.
I can construct and interpret grouped frequency diagrams.

Review questions

1 Josh asked ten people: 'Who are your favourite singers?' Here are his results.

Lady Gaga   Bruno Mars   Lady Gaga   Jared Leto   Emeli Sande

Jared Leto   Emeli Sande   Lady Gaga   Adam Lambert   Lady Gaga

  a Is it possible to work out the mean of these results?
      Explain your answer.

  b Is it possible to work out the mode of these results?
      Explain your answer.

2 a David played eight games in a competition. In each of six games, he scored 10 points. In the other two games he scored no points. What was David's mean score over the eight games?

  b Jess played three games. Her mean score was 4 points. Her range was 3 points. What possible points did Jess score in each of her three games?

  c Andy played five games. His mean score was 5 points. His range was 6 points. What individual points might Andy have scored in each of his five games?

3 a The modal mass of 12 boys is 65 kg. What can you say about their total mass?

  b The median mass of 8 girls is 56 kg. What can you say about their total mass?

  c The mean mass of 10 people is 79.6 kg. What can you say about their total mass?

4 The cost of a clown's mask varies with the height, as shown in the table.

Height (cm)	15	20	25	30	35
Cost (£)	10	18	27	38	51

  a Plot these points on a graph. Join up the points.

  b From your graph, estimate:

     i the cost of a mask of height 28 cm   **ii** the height of a mask that cost £20.

5   The table shows the lengths of time Philip waited for a bus in one month.

Time, T (minutes)	Frequency
$0 < T \le 5$	3
$5 < T \le 10$	10
$10 < T \le 15$	4
$15 < T \le 20$	3

a   One Friday he waited 15 minutes for his bus. In which class group is this time recorded?

b   What is the modal class?

c   What is the probability of his waiting ten minutes or less for a bus?

6   A basketball team was not scoring enough points. The coach wanted to bring in a new player, who might increase the scores.

There were two players to choose from. These were their scoring records for the last six seasons.

Gary	45	36	48	57	42	45
Mark	63	18	21	24	63	27

a   Give a reason to choose Gary.

b   Give a reason to choose Mark.

c   Which player would you choose if you were the coach? Explain your choice.

7   Kathryn was a piano teacher. Over six months she monitored how many lessons she was giving each week. These are her results.

12, 16, 27, 9, 13, 16, 10, 21, 19, 17, 24, 8, 7, 12, 19, 23, 11, 25, 21, 17, 16, 20, 14, 9, 8, 17

a   Put this data into a suitable grouped frequency table.

b   Construct a frequency diagram from the data.

c   What were Kathryn's modal class numbers?

(PS) 8   These are three integers. Their range is 4.

8, $x + 3$, 10

a   What is the value of x, when the mode is 8?

b   What is the value of x, when the median is 10?

c   What is the value of x, when the mean is 8?

(PS) 9   a   What is the mean of the perimeters of these three triangles?

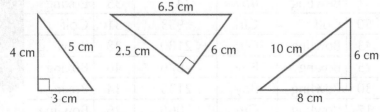

b   What is the mean area of the triangles?

Problem solving
Technology questionnaire

Lucy wrote a questionnaire for her year group.

She asked 25 boys and 25 girls these questions.

a How many downloaded music tracks do you have?

b How many CDs do you own?

c Which is your favourite Wii sporting game?

This is a summary of her results.

Boy/girl	Downloads	CDs	Wii game	Boy/girl	Downloads	CDs	Wii game
Boy	1824	12	Bowling	Girl	2012	17	Bowling
Girl	1632	17	Bowling	Boy	3008	32	Golf
Boy	2187	27	Boxing	Girl	3654	43	Bowling
Girl	2562	34	Bowling	Girl	2219	26	Golf
Girl	2193	32	Bowling	Boy	2187	32	Boxing
Boy	1086	29	Boxing	Boy	3280	44	Boxing
Boy	1243	43	Golf	Boy	3098	53	Boxing
Boy	1786	41	Boxing	Girl	2021	45	Bowling
Girl	1329	23	Bowling	Girl	783	36	Bowling
Girl	958	16	Golf	Boy	1086	28	Boxing
Boy	1982	26	Boxing	Boy	762	16	Tennis
Girl	1090	37	Bowling	Girl	542	19	Tennis
Girl	1824	24	Bowling	Boy	781	20	Golf
Boy	1655	16	Golf	Girl	2005	31	Bowling
Boy	2311	33	Boxing	Girl	1019	40	Bowling
Girl	4044	45	Bowling	Boy	2109	43	Bowling
Boy	4109	52	Tennis	Girl	3218	55	Golf
Girl	3381	37	Tennis	Girl	1894	23	Bowling
Girl	564	13	Bowling	Boy	2015	27	Baseball
Girl	2034	31	Bowling	Boy	2415	35	Boxing
Boy	4298	50	Golf	Girl	953	31	Golf
Boy	3026	43	Bowling	Girl	2180	28	Bowling
Boy	2051	25	Boxing	Boy	2876	46	Boxing
Girl	1980	30	Bowling	Boy	2133	34	Boxing
Boy	1950	17	Bowling	Girl	1066	26	Bowling

1 Draw and complete tally charts for the favourite Wii game for boys and for girls.

2 From your tally charts draw pie charts for:

   a boys   b girls.

3 Draw up a grouped frequency table for the number of downloaded tracks for:

   a boys   b girls.

4 Draw a frequency chart to help you compare any differences between the numbers of downloaded tracks for girls and boys.

5 What is the probability that a pupil chosen from this sample at random:

   a is a boy

   b has over 2000 downloaded tracks

   c owns more than 30 CDs

   d has tennis as their favourite Wii game?

Glossary

π The result of dividing the circumference of a circle by its diameter, represented by the Greek letter pi.

algebraic expression An expression that contains numbers, variables and one or more arithmetic operations.

allied angles Interior angles that are formed on the same side of a transversal that cuts a pair of parallel lines; they add up to 180°.

alternate angles Angles that lie on either side of a transversal that cuts a pair of parallel lines; the transversal forms two pairs of alternate angles, and the angles in each pair are equal.

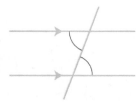

angle bisector A line or line segment that divides an angle into two equal parts.

angle of rotation The angle through which an object is rotated, to form the image.

approximate Work out a value that is close but not exactly equal to another value, and can be used to give an idea of the size of the value; for example, a journey taking 58 minutes may be described as 'taking approximately an hour'; the sign ≈ means 'is approximately equal to'.

arc Part of the circumference of a circle.

average speed The result of dividing the total distance travelled by the total time taken for a journey.

base One of the sides of a 2D shape, usually the one drawn first, or shown at the bottom of the shape.

bisect Cut exactly in half.

block graph A graph in which blocks, bars or columns represent data values; another name for a bar chart.

centre The point inside the circle that is the same distance from every point on the circumference.

centre of enlargement The point, inside or outside the object, on which the enlargement is centred; the point from which the enlargement of an object is measured.

centre of rotation The point about which an object or shape is rotated.

chord A straight line joining two points on the circumference of a circle; a diameter is a chord that passes through the centre of the circle.

circumference The perimeter of a circle; every point on the circumference is the same distance from the centre, and this distance is the radius.

class A small range of values within a large set of data, treated as one group of values.

coefficient A number written in front of a variable in an algebraic term; for example, in $8x$, 8 is the coefficient of x.

common factor A factor that divides exactly into two or more numbers; 2 is a common factor of 6, 8 and 10.

compound shape A shape made from two or more simpler shapes; for example, a floor plan could be made from a square and a rectangle joined together.

congruent Exactly the same shape and size.

congruent triangles Triangles that are exactly the same shape and size; conditions are:

 ASA – two angles and the included side

 SAS – two sides and the included angle

 SSS – three sides

 RHS – right angle, hypotenuse and side

constant A value that does not change; in the equation $y = 3x + 6$, the values of x and y may change, but 6 is a constant.

constant term A term that has a fixed value; in the equation $y = 3x + 6$, the values of x and y may change, but 6 is a constant term.

construction An accurate drawing, made with ruler and compasses.

continuous Having an infinite number of values, with no break; height is a continuous measurement.

correlation A relationship in which two or more things affect each other.

corresponding angles Angles that lie on the same side of a pair of parallel lines cut by a transversal; the transversal forms four pairs of corresponding angles, and the angles in each pair are equal.

cube **1** In geometry, a 3D shape with six square faces, eight vertices and 12 edges.

2 In number and algebra, the result of multiplying a number or expression raised to the power of three: n^3 is read as 'n cubed' or 'n to the power of three':

for example: 2^3 is the cube of 2 and $(2 \times 2 \times 2) = 8$.

cube root For a given number, a, the cube root is the number b, where $a = b^3$; for example, the cube root of 8 is 2 since $2^3 = 8$. The cube root of 8 is recorded as $\sqrt[3]{8} = 2$.

cuboid A 3D shape with six rectangular faces, eight vertices and 12 edges; opposite faces are identical to each other.

decimal place The position, after the decimal point, of a digit in a decimal number; for example, in 0.025, 5 is in the third decimal place. Also, the number of digits to the right of the decimal point in a decimal number; for example, 3.142 is a number given correct to three decimal places (3 dp).

diameter A straight line joining two points on the circumference of a circle, and passing through the centre.

direct proportion A relationship in which one variable increases or decreases at the same rate as another; in the formula $y = 12x$, x and y are in direct proportion.

direction of rotation The direction, clockwise or anticlockwise, in which an object is rotated to form an image.

discrete Able to take only certain values, such as numbers of children in a family.

distance–time graph A graph showing the distance travelled (vertical axis) against the time taken (horizontal axis), for a journey.

divisible Able to be divided exactly; 6 is divisible by 2 and by 3, 8 is divisible by 2 and 4 but not by 3.

divisor The number being divided into another number; in $810 \div 5$, 5 is the divisor.

enlargement A transformation in which the object is enlarged to form an image.

equally likely   When the probabilities of two or more outcomes are equal; for example, when a fair six-sided dice is thrown, the outcomes 6 and 2 are equally likely with probabilities of $\frac{1}{6}$.

equivalent   The same, equal in value.

estimate   **1** State or guess a value, based on experience or what you already know.

2 A rough or approximate answer.

event   Something that happens, such as the toss of a coin, the throw of a dice or a football match.

expand   To expand a term with brackets, you multiply everything inside the brackets by the value in front of the brackets.

experimental probability   The probability found by trial or experiment; an estimate of the true probability.

factor   A number that divides exactly into another number, without leaving a remainder.

factor tree   A diagrammatic method of finding the prime factors of a number, by dividing it by its prime factors.

Fibonacci sequence   A sequence of numbers in which the third and subsequent terms are formed by adding the two previous terms: 1, 1, 2, 3, 5, 8, …

finite   Within a countable limit.

flow diagram   A diagram that shows a sequence of operations or activities.

formula   A mathematical rule, using numbers and letters, that shows a relationship between variables; for example, the conversion formula from temperatures in Fahrenheit to temperatures in Celsius is: $C = \frac{5}{9}(F - 32)$.

frequency   The number of times a particular item appears in a set of data.

frequency diagram   A diagram, such as a pie chart or bar chart, that shows the frequencies of data in a set.

frequency table   A table showing data values, or ranges of data values, and the numbers of times that they occur in a survey or trial.

generate   Create or form.

geometric properties   The properties of a 2D or 3D shape that describe it completely.

gradient   The slope of a line between two or more points, calculated as the vertical difference divided by the horizontal difference.

graph   A diagram showing the relation between certain sets of numbers or quantities by means of a series of values or points plotted on a set of axes.

grouped frequency table   A table showing data grouped into classes.

hectare   A metric unit of area, equal to 10 000 m².

height   The vertical distance, from bottom to top, of a 2D or 3D shape.

highest common factor (HCF)   The largest number that is a factor common to two or more other numbers.

index   Power; in 3^4, 4 is the index.

index form   A number that is expressed as another number raised to a power is in index form.

infinite   Having no limit or end point, such as the set of integers.

integer   Whole number, may be positive or negative, including zero.

intersection   The 'overlap', the set of data values that occur in both of two sets.

inverse operation An operation that reverses the effect of another operation; for example, addition is the inverse of subtraction, division is the inverse of multiplication.

inverse proportion A relationship between two variables in which as one value increases, the other decreases; in the formula $xy = 12$, x and y are in inverse proportion.

like terms Terms in which the variables are identical, but have different coefficients; for example, $2ax$ and $5ax$ are like terms but $5xy$ and $7y$ are not. Like terms can be combined by adding their numerical coefficients so $2ax + 5ax = 7ax$.

line graph A chart that shows how data changes, by means of points joined by straight lines.

linear equation An equation such as $y = 4x - 7$, that will produce a straight-line graph.

lowest common multiple (LCM) The lowest number that is a multiple of two or more numbers; 12 is the lowest common multiple of 2, 3, 4 and 6.

manipulate Rearrange an algebraic statement to put it into a useful form.

map ratio The ratio of the distance on a map to the distance it represents on the ground.

multiple A number that results from multiplying one number by another; multiples appear in the multiplication table of a number.

multiplier A number that is used to find a percentage; the multiplier for 75% is 0.75 and the multiplier for 105% is 1.05.

multiply out To multiply out brackets, you multiply everything inside the brackets by the number or term outside the brackets.

mutually exclusive Unable to occur at the same time.

negative number A number that is less than zero.

nth term An expression in terms of n; it allows you to find any term in a sequence, without having to use a term-to-term rule.

outcome A result of a trial or event.

parallelogram A quadrilateral with two pairs of parallel sides; the opposite sides are equal in length.

percentage A number written as a fraction with 100 parts, but instead of writing it as a fraction out of 100, you write the symbol % at the end, so $\frac{50}{100}$ is written as 50%.

perpendicular bisector A line that divides a given line exactly in half, passing through the midpoint at right angles to it.

perpendicular height The distance between the base of a 2D shape and its topmost point or vertex, measured at right angles to the base.

pie chart A circular graph divided into sectors that are proportional to the size of the quantities represented.

positive number A number that is greater than zero.

power How many times you use a number or expression in a calculation; it is written as a small, raised number; for example, 2^2 is 2 multiplied by itself, $2^2 = 2 \times 2 = 4$ and 4^3 is $4 \times 4 \times 4 = 64$.

prime factor A factor that is a prime number; 2 and 3 are the prime factors of 6.

prime number A number that has exactly two factors, itself and 1; the number 1 is not a prime number, as it only has one factor.

probability The measure of how likely an outcome of an event is to occur. All probabilities have values in the range from 0 to 1.

probability scale A scale or number line, from 0 to 1, sometimes labelled with impossible, unlikely, even chance, etc., to show the likelihood of an outcome of an event occurring. Possible outcomes may be marked along the scale as fractions or decimals.

proportional In proportion.

quadratic Containing a term with a squared variable, such as $y = 2x^2 + 4$.

radius The shortest distance between the centre of a circle and its circumference.

ray A straight line, drawn through two points, used to enlarge a shape.

rotation A turn about a central point, called the centre of rotation.

round Approximate according to a given condition, such as a number of decimal places or significant figures.

sample space The set of all possible outcomes for an event or trial.

scale The ratio of the length on the image to the length on the object.

scale drawing A drawing that represents something much larger or much smaller, in which the lengths on the image are in direct proportion to the lengths on the object.

scale factor The ratio of the distance on the image to the distance it represents on the object; the number that tells you how much a shape is to be enlarged.

scaling A method used in drawing statistical diagrams, such as pie charts; data values are multiplied or divided by the same number, so that they can be represented proportionally in a diagram.

scatter graph A graphical representation showing whether there is a relationship between two sets of data.

sector A region of a circle, like a slice of a pie, enclosed by an arc and two radii.

segment A part of a circle cut off by a chord.

semicircle Half of a circle, based on a diameter.

set A group of values, such as the ages of pupils in a class.

significant In a number, the digits that give an approximation of its value are significant.

significant figure In the number 12 068, 1 is the first and most significant figure and 8 is the fifth and least significant figure. In 0.246 the first and most significant figure is 2. Zeros at the beginning of a number are not significant figures.

square Multiply a number by itself.

square root For a given number, a, the square root is the number b, where $a = b^2$; for example, a square root of 25 is 5 since $5^2 = 25$. The square root of 25 is recorded as $\sqrt{25} = 5$. Note that a positive number has a negative square root, as well as a positive square root; for example, $(-5)^2 = 25$ so it is also true that $\sqrt{25} = -5$.

standard form A way of writing very small or very large numbers in the form $A \times 10^n$, where A is a positive number, greater than or equal to 1 but less than 10 ($1 \leq A < 10$) and n is an integer.

subject In a formula, the subject is on its own, on the left-hand side of the equals sign.

surface area The total area of all of the surfaces of a 3D shape.

tangent A straight line that touches a circle just once.

term **1** A part of an expression, equation or formula. Terms are usually separated by + and − signs.

2 A number in a sequence or pattern.

total frequency The result of adding together all of the frequencies in a date set.

transformation A change to a geometric 2D shape, such as a translation, rotation, reflection or enlargement.

translate Move in a straight line, horizontally, vertically or diagonally.

translation A movement along, up or diagonally on a coordinate grid.

transversal A straight line that cuts two or more parallel lines.

trapezium A quadrilateral in which only one pair of opposite sides are parallel but unequal in length. In an isosceles trapezium, the other two sides are the same length as each other.

union The set of data items that occur in one, the other or all of two or more data sets.

unit fraction A fraction with 1 as the numerator.

variable A quantity that may take many values.

Venn diagram A diagram that uses closed loops, such as circles or ellipses, to represent the data items in sets.

Index

William Collins's dream of knowledge for all began with the publication of his first book in 1819. A self-educated mill worker, he not only enriched millions of lives, but also founded a flourishing publishing house. Today, staying true to this spirit, Collins books are packed with inspiration, innovation and practical expertise. They place you at the centre of a world of possibility and give you exactly what you need to explore it.

Collins. Freedom to teach.

Published by Collins
An imprint of HarperCollins*Publishers*
The News Building
1 London Bridge Street
London
SE1 9GF

HarperCollins *Publishers*
Macken House,
39/40 Mayor Street Upper,
Dublin 1,
D01 C9W8
Ireland

Browse the complete Collins catalogue at
www.collins.co.uk

12

ISBN-13 978-0-00-753775-4

Kevin Evans, Keith Gordon, Trevor Senior, Brian Speed and Chris Pearce assert their moral rights to be identified as the authors of this work.

British Library Cataloguing in Publication Data
A catalogue record for this publication is available from the British Library.

Commissioned by Katie Sergeant
Project managed by Elektra Media Ltd
Development edited by Lindsey Charles
Copy-edited and proofread by Joan Miller
Edited by Helen Marsden
Proofread by Amanda Dickson
Illustrations by Ann Paganuzzi, Nigel Jordan and Tony Wilkins
Typeset by Jouve India Private Limited
Cover design by Angela English
Printed and bound in the UK by Ashford Colour Press Ltd

Acknowledgements

The publishers wish to thank the following for permission to reproduce photographs. Every effort has been made to trace copyright holders and to obtain their permission for the use of copyright materials. The publishers will gladly receive any information enabling them to rectify any error or omission at the first opportunity.

(t = top, c = centre, b = bottom, r = right, l = left)

Cover gyn9037/Shutterstock, p 6 stocksolutions/Shutterstock, p 15 Rafal Olechowski/Shutterstock, p 24–25 Pefkos/Shutterstock, p 26 Brian A Jackson/Shutterstock, p 48–49 Hallgerd/Shutterstock, p 50 Paolo Nespoli/ESA/NASA via Getty Images, p 68–69 clearviewstock/Shutterstock, p 70 Lightspring/Shutterstock, p 82–83 bikeriderlondon/Shutterstock, p 84 Filip Fuxa/Shutterstock, p 98–99 Maceofoto/Shutterstock, p 100 jean morrison/Shutterstock, p 118–119 tavi/Shutterstock, p 120 design36/Shutterstock, p 134–135 Adrian Reynolds/Shutterstock, p 136 Suppakij1017/Shutterstock, p 151 antb/Shutterstock, p 152–153 aquatic creature/Shutterstock, p 154 Sergey Nivens/Shutterstock, p 170–171 Dean Mouhtaropoulos/Getty Images, p 172 bekulnis/Shutterstock, p 186–187 WitR/Shutterstock, p 188 *Mona Lisa*, c.1503–6 (oil on panel), Vinci, Leonardo da (1452–1519)/Louvre, Paris, France/Giraudon/The Bridgeman Art Library, p 193 Ann Paganuzzi, p 206–207 Yeko Photo Studio/Shutterstock, p 208 andersphoto/Shutterstock, p 224–225 wavebreakmedia/Shutterstock, p 226 *The Proportions of the human figure (after Vitruvius)*, c.1492 (pen & ink on paper), Vinci, Leonardo da (1452–1519)/Galleria dell' Accademia, Venice, Italy/The Bridgeman Art Library, p 244–245 Taina Sohlman/Shutterstock, p 246 AlexanderZam/Shutterstock, p 260–261 ArtThailand/Shutterstock, p 262 SergeyDV/Shutterstock, p 272 Correcaminos112/Shutterstock, p 273 pjcross/Shutterstock, p 276–277 Chukcha/Shutterstock, p 278 Robert Nyholm/Shutterstock, p 282 Nicholas Piccillo/Shutterstock, p 286 mart/Shutterstock, p 292–293 nmedia/Shutterstock.